G000097496

Environmental Issues and Business

Wyn Grant
PAIS
University of Warwick

Environmental Issues and Business

Implications of a Changing Agenda

Sally Eden

Middlesex University, Enfield, UK

JOHN WILEY & SONS

Chichester • New York • Brisbane • Toronto • Singapore

Published 1996 by John Wiley & Sons Ltd,
Baffins Lane, Chichester,
West Sussex PO19 1UD, England

National   01243 779777
International (+44) 1243 779777

e-mail (for orders and customer service enquiries): cs-books@wiley.co.uk

Visit our Home Page on http://www.wiley.co.uk
or http://www.wiley.com

Other Wiley Editorial Offices

John Wiley & Sons, Inc., 605 Third Avenue,
New York, NY 10158-0012, USA

Jacaranda Wiley Ltd, 33 Park Road, Milton,
Queensland 4064, Australia

John Wiley & Sons (Canada) Ltd, 22 Worcester Road,
Rexdale, Ontario M9W 1L1, Canada

John Wiley & Sons (Asia) Pte Ltd, 2 Clementi Loop #02-01,
Jin Xing Distripark, Singapore 129809

Library of Congress Cataloging-in-Publication Data
Eden, Sally.
    Environmental issues and business : implications of a changing agenda / Sally Eden.
        p.   cm.
    Includes bibliographical references and index.
    ISBN 0-471-94872-1
    1. Industrial management—Environmental aspects—Great Britain.   2. Social responsibility of business—Great Britain.   3. Environmental policy—Great Britain.   I. Title.
    HD30.255.E34   1996
    658.4'08—dc20   96-16207
                                                                CIP
British Library Cataloguing in Publication Data

A catalogue record for this book is available from the British Library

ISBN 0-471-94872-1

Typeset in 10/12pt Ehrhardt from author's disks by Mayhew Typesetting, Rhayader, Powys
Printed and bound in Great Britain by Biddles Ltd, Guildford and King's Lynn

This book is printed on acid-free paper responsibly manufactured from sustainable forestation, for which at least two trees are planted for each one used for paper production.

Contents

Preface

When I began to research business responses to environmental issues, I found few integrated references on which to draw. There seemed, therefore, a need for a book which addressed the multitude of questions raised by such developments in the 1990s. Of course, since then, other authors have similarly recognised this need and produced books on business and environmental strategies, technologies and change. However, I still feel that many of these books address the questions from a stance firmly rooted within the traditional business management paradigm, whereas the few additional environmentalist texts about business tend to propound polemical criticism rather than analysis. Whilst the yawning gap of books on the topic has been remedied to an extent, there remain few texts which approach it with the cautious gait and cultural critique of a social science perspective.

I have therefore chosen to take a critical look at business and the environment from the social sciences. This had to be an interdisciplinary exercise because the sparse yet rapidly expanding work on the topic reaches across geography, sociology, economics, political science and management studies. From this range of material, I have tried to select the key themes and to explore their current status and support. I have focused particularly on the implications of business activity for redefining the environmental agenda, in political, social and policy-related terms. As well as merely reacting to criticism, business is proactively seeking greater and greater input to the environmental debate. To me, it is here that we may see the most significant effects of business interest in environmental issues, and it is therefore here that we must direct our closest attention. In my examples, I have necessarily concentrated on European and North American developments, because these have been prominent and 'pioneering' within the business community, founded as they are on the greater resources of the large corporations based in these areas. I have also focused on the UK, which is where my own research into business and the environment has been conducted. My coverage will necessarily be incomplete and I can only hope that this prompts those with greater experience in other countries to fill the remaining gaps in the literature through their own publications. It seems that the entire field is continually undergoing evolution and multiplication as political, business and academic interest in environmental issues builds.

This book is intended for those who are entering this field from different research standpoints and from different disciplines. I hope that it will prove useful to those who have concentrated on related subjects in the fields of business or environmental studies and who wish to examine the crossovers and interconnections. Doubtless, some will feel that I have been too harsh with business and some that I have been too kind. I hope at least that my treatment will contribute to the growing debates about business and environmental issues and the implications for the wider environmental agenda.

Some of the material incorporated in the text has already been published elsewhere in some form, particularly: parts of Chapters 3 and 4 in Eden, S.E. (1993) 'Constructing environmental responsibility: perceptions from retail business' *Geoforum* **24**, 4, 411–21 and Eden, S.E. (1995) 'Business, trust and environmental information: perceptions from consumers and retailers' *Business Strategy and the Environment* **3**, 4, 1–8 and parts of Chapter 6 in Eden, S.E. (1994) 'Using sustainable development: the business case' *Global Environmental Change* **4**, 2, 160–7. The 'learning curve of greening' in Figure 3.1 is reproduced with permission from John Elkington, Peter Knight with Julia Hailes (1991) *The Green Business Guide* (London, Victor Gollancz), p.230 and Axel Scheffler's illustrations in Figures 1.2 and 4.1 have been reproduced with permission from the pages of *Resurgence* (153, p. 8 and 143, p. 20 respectively). I approached a number of advertisers for permission to reproduce their examples of environmental advertising as illustrations for my arguments in Chapter 3. Unfortunately, I could not obtain any permissions and have had to content myself with verbal descriptions.

Much of the material in Chapters 5 and 6 was gathered with the aid of a grant from the Nuffield Foundation, UK, and I would like to thank them for their support. During the course of that grant, I interviewed a number of business representatives, some of whom are quoted in the text, and I would like to thank them all for providing me with their time and their views of business activity. A number of people have been kind enough to comment on earlier drafts of certain chapters, namely Andy Gouldson, Andy Cumbers, Dave North, Dave Smallbone and particularly Martin Purvis, whose detailed and stimulating comments caused me to rethink some complex questions. I would like to thank them for their help whilst holding them in no way responsible for my interpretation of their comments in the final text. I would also like to thank the Technical Unit in GEM at Middlesex University, principally Ailsa Farquhar and Steve Chilton, who provided the illustrations.

So far, I have lived in five different houses whilst writing this book and I will be moving again before it is published. I have to thank both my parents for spending considerable time and effort to ensure that things went smoothly and especially Antony, for helping me to keep it together despite all the moves and for his support throughout my work.

List of abbreviations

ACBE	Advisory Committee on Business and the Environment
ASA	Advertising Standards Authority (UK)
BATNEEC	best available technique not exceeding excessive cost
BCSD	Business Council for Sustainable Development
BPEO	best practicable environmental option
BS 7750	British Standard 7750 (Environmental Management Systems)
CBI	Confederation of British Industry
CEFIC	European chemical industry federation
CEO	Chief executive officer
CERCLA	Comprehensive Environmental Response, Compensation and Liability Act (a.k.a. Superfund, USA)
CFC	Chlorofluorocarbon
CIA	Chemical Industries Association (UK)
COPAC	Consortium of the Packaging Chain (UK)
DG	Directorate General (EC)
DOE	Department of the Environment (UK)
DTI	Department of Trade and Industry (UK)
EC	European Commission
EMAS	Eco-Management and Audit Scheme (set up by EC)
EMS	Environmental management systems
EPA	Environmental Protection Agency (USA)
EPA 1990	Environmental Protection Act 1990 (UK)
EU	European Union
HMIP	Her Majesty's Inspectorate of Pollution (UK)
ICC	International Chamber of Commerce
IISD	International Institute for Sustainable Development
INCPEN	Industry Council for Packaging and the Environment (UK)
IOD	Institute of Directors (UK)
IPC	Integrated Pollution Control (UK)
IPPC	Integrated Pollution Prevention and Control (EC)
n.d.	no date
NGO	Non-governmental organisation

NRA	National Rivers Authority (UK)
PR	Public relations
PRG	Producer Responsibility Industry Group (UK)
SARA	Superfund Amendments and Reauthorization Act (USA)
SME	Small to medium-sized enterprise
UNCED	United Nations Conference on Environment and Development
UNEP	United Nations Environment Programme
WCED	World Commission on Environment and Development
WICE	World Industry Council for the Environment (ICC)
WICEM	World Industry Conference on Environmental Management

Environmental issues and business: an introduction

The relationship between business and environmental issues is a complex one, which both influences and is influenced by changes to the environmental agenda. In the late 1980s in western economies, environmental issues became much more a part of the business agenda than before. Previously, business had been cast as the environmental villain, preventing it from developing a positive response. However, in the 1980s environmentalism became less radical, reflecting the neo-conservative politics of the time, dominated by 'a widespread acceptance of the supremacy of the market mechanism as the only sound basis for an economy' in the wake of the disintegration of the Soviet Union and changes in Eastern Europe (Grant 1993, p.1). Since then, business has approached environmental issues in a more positive way as 'green business', using economic arguments and market mechanisms to offer solutions to environmental damage. Thus business has recast itself against type as the environmental saviour. Its critics in turn have dismissed such a 'green' conversion as false, and argued that the very idea of 'green business' is a contradiction in terms and a misleading representation of merely superficial change.

This book considers both sets of arguments by outlining the activities of business in response to environmental issues and evaluating the corresponding critiques of these activities. Because the notion of 'green business' has an impact on politics, economics, societies and business cultures, as well as the environment, I shall be taking an interdisciplinary perspective to this task. I shall also be concentrating on developments in Europe, especially the UK, and North America because it is here that public concern, governmental attention and not least the organisation of business have caused environmental issues to make their greatest impact. It is also here that the large corporations which are the prime movers for environmental activity in business fields are mainly based.

The subsequent chapters look at different ways in which business has responded to environmental issues. First, business has considered its own operations and their environmental impacts within single companies and across sectors and supply chains, leading to changes in management, organisational structures and production and distribution operations (Chapter 3). Secondly, business has looked at its communication strategies in the light of environmental concerns shown by various publics, leading to changes in advertising, marketing, policy statements and reporting (Chapter 4). Thirdly, it has changed its political and regulatory activities, incorporating environmental considerations into lobbying, political representation and its relationships with regulators and non-governmental organisations (NGOs) (Chapters 5 and 6). In order to frame this variety of activity, I have sought to deal in the earlier chapters (2, 3 and 4) with its more reactive forms, activities which have developed in response to pressures from markets, NGOs, policy, legislation and other businesses. Later chapters (5 and 6) deal more with proactive forms through policy and regulatory input and self-regulation. Of course, this is a rather artificial distinction, because reaction to policy can lead to proaction or proaction may only be one jump ahead of an inevitable policy innovation. I shall note throughout the book how closely proaction and reaction are intertwined and attempt to set both within the larger sociopolitical and economic context. Still, the proaction–reaction distinction is a useful structuring device for the broad set of topics with which this book is concerned.

This chapter serves as an introduction to and historical appreciation of a number of these topics. I shall begin by discussing the roots of business interest in the environment and move on to the rationales for business environmental interest and the arguments surrounding the connections between business and the environment.

THE DEVELOPMENT OF ENVIRONMENTAL INTEREST

In order to understand how business interest in the environment developed, we need to appreciate the long history of its development, especially public and policy attention to the environment more generally. Environmental issues have been connected to industry since the earliest pollution incidents. For example, in 1556, a writer claimed that mining yielded a 'greater detriment' to the environment than it produced value in minerals (Elkington and Burke 1987, p.31). But it was during the industrial expansion from the late eighteenth century that industrial regulation developed to deal with increasingly visible environmental damage. For example, in Lancashire at this time:

> what was to be the most appalling town of all—St Helens—was just beginning to define its surroundings [through its glassworks and copperworks] . . . The atmosphere was being poisoned, every green thing blighted, and every stream fouled with chemical fumes and waste. (Hoskins 1970, p.222)

Roused by the public health effects of such pollution, public concern led to the first real piece of environmental legislation, the Alkali Acts in the UK in the 1860s, developed to deal with the pollution from alkali works (Ball and Bell 1991; Wohl 1983, p.228). This was followed by early environmental campaigning from the latter half of the nineteenth century in western economies, concentrating first on preservation of the natural environment. The first UK environmental group, the Commons, Open Spaces and Footpaths Preservation Society established in 1865, was followed by the Royal Society for the Protection of Birds in 1889 and the National Trust in 1895, the latter focusing on the preservation of historic buildings and sites as well as landscapes and environments (Lowe and Goyder 1983). In the USA, the wilderness movement was underway, with the first National Park established at Yellowstone in 1872 (McEvoy 1972), and National Parks were also attracting support in Australia with the Royal National Park at Sydney in 1879 (Eckersley 1992) and eventually in Europe and the UK (MacEwen and MacEwen 1987).

By the early twentieth century, the neonatal environmental lobby had begun to expand as people increasingly turned to the environment for physical recreation. Amenity associations sprang up, such as the Ramblers Association (UK, 1925), and pressure groups were established with interests different to traditional preservation ones, e.g. the Soil Association which focused on farming practices (UK, 1946). The two foci for environmentalism at this time were the older strand of 'preservationism', exemplified by the protection of 'wilderness' areas, and the newer strand of 'conservationism', which focused on the utility of the environment and improved technical and resource management in long-inhabited areas.[1] After the Second World War, the new town and country planning movement in the UK refocused attention on urban environments, and the industry within them, and was complemented by various legislative moves to preserve greenbelt land and to control domestic and industrial pollution (Ball and Bell 1994).

But it was in the late 1960s that environmental interest really reached public consciousness. The social and political climate was changing as people became far more publicly critical of the state and its agencies. Movements for civil rights, student protests and anti-Vietnam campaigns all energised the political culture across Europe and North America. New ways of protesting and living were advocated and public figures and their policies, as well as industries, were increasingly exposed to the rigours of public legitimation. Morrison et al. (1972) ascribed these cultural developments to a 'participation orientation' coupled with a 'media-led euphoria' about social and environmental issues. People became involved in campaigning at the grass-roots level, rather than the more elitist, hierarchical environmental campaigning that had previously been predominant. Hence, this period saw the birth of what we now think of as environmentalism (O'Riordan 1976, p.51) and large national and international pressure groups emerged which had a similar orientation towards more radical and grass-roots activities. Friends of the Earth was born out of disagreements within the more traditional Sierra Club in the USA and Greenpeace out of protests against nuclear testing in the Pacific (McCormick 1989; Yearley 1991; and see Table 1.1). This 'organised environmentalism' was paralleled by 'institutional environmentalism' (Buttel and Larson 1980) as old institutions restructured to take account of environmental issues and new ones were instigated, such as the USA's Environmental Protection Agency and

Table 1.1   1970s environmental groups and initiatives

Date	Environmental group or initiative
1969	Friends of the Earth (USA)
1970	Department of the Environment (UK)
1970	Environmental Protection Agency (USA)
1970	Earth Day, 22 April (USA)
1972	United Nations Conference on the Human Environment, Stockholm
1972	First green party established (Values Party in New Zealand)
1973	People established (later the Ecology Party and then the Green Party, UK)

Source: Cotgrove (1982); Rüdig and Lowe (1986).

the UK's Department of the Environment. The mass mobilisation of the 1960s and early 1970s (Buttel 1986) radicalised environmental protest and emphasised public support, involvement and collective action.

Environmentalism in this period was not conducive to support for business environmental changes. In the 1950s, people had seen industry, like science, in a positive light as the engine of economic growth and the provider of better living standards, better and more labour-saving technology and a brighter future (a light which industry of course encouraged through advertising). But the critique of the state in the 1960s was coupled with critiques of the role of business and of science (especially atomic science), which increasingly cast business as the villain of environmental damage. This loss of public trust and therefore legitimation has never been fully restored, especially for the chemical and energy sectors (see Chapter 5; Eden 1995). It was further fuelled by the wider reporting of damaging incidents in the 1960s and 1970s, such as the *Torrey Canyon* oil spill off the Isles of Scilly in south-west Britain. Cleaning up the resultant pollution of the Cornish coast cost £6 million and led to the later establishment of the Royal Commission on Environmental Pollution in the UK (McCormick 1995). Environmentalists publicised critiques of the ethos of growth that ran counter to the foundations of business thinking. They not only criticised the environmental record of industry but also business's commitment to growth as an objective in itself as socially and economically irresponsible.

Consequently, by 1970 business, and especially manufacturing industry, could no longer promote itself successfully as the solution to environmental problems in a positive way. It had to develop a defensive stance aimed at countering images of itself as the environmental villain (Hoffman 1994). The 1972 *Limits to Growth* report (Meadows *et al.* 1972), produced for a business group, the Club of Rome, intimated the need for business to respond energetically to environmental concerns, especially over growth, and prompted much discussion about whether natural resources would last in the face of predicted increases in global demand. Specific foci of dispute in this period were the possibilities of technological adaptation, the relative nature of physical (environmental) limits to growth, the problems of predicting growth (see O'Riordan 1981; Sandbach 1978; Harvey 1974) and the accusation that it was merely 'doom-mongering' to suggest that the industrial system would lead to the collapse of society in its current form (Maddox 1972).

Overall, the early 1970s were characterised by fairly negative interpretations of environmental (and population) crises. These debates reoriented the agenda from the 1960s' political activism and re-established the importance of economic constraints. Environmental interest shifted from a 'participation' orientation which excluded and criticised business, to a 'survival' orientation (Eckersley 1992). The latter brought business into the disputes as a key agent of survival or collapse. The subsequent energy crisis of the mid-1970s reinforced the shift from grass-roots radicals to 'growthists'—business people, politicians and their supporters who were committed to growth (Buttel 1986). The conflicts between growthists and zero-growth advocates in this period led to a diversity of perspectives and pressure groups being established. Middle-range ideas surfaced, such as appropriate technology, soft energy and sustainable development—the latter became very significant by the end of the 1980s as a means of bridging the growth and no-growth arguments (Chapter 6).

The character of environmentalism in the 1960s and early 1970s did not encourage the development of environmental activity by business, because it was grass-roots oriented, collectivist and interested in ideas of zero-growth which business could not accommodate within its ideology. However, the institutionalisation of environmental issues was recognised by business and its commentators and had two consequences. First, and related to the emphasis on public participation, there was much discussion about how business might adopt social responsibility in the early 1970s (Gray et al. 1987) prompting antagonists to emerge (e.g. Drotning 1972). This argument was based on the notion that a business's powerful position in society required it to act responsibly towards the community and individuals as well as to its shareholders (but this was not the consensus, cf. Friedman 1988; see Chapter 2). These ideas centred around employee rights, information provision, cooperation with local authorities and unions and dialogue with the community, all disclosed through 'social auditing' of these activities. They were also implicitly linked to ensuring legitimation for companies and business as whole, to remedy the distrust engendered since the 1960s.

A second consequence of environmentalist activity was the expansion of legislation coupled with the higher profile of its enforcement through stronger regulation. Until the 1970s, legislation had been medium-specific, with separate regulations for water, air and land pollution. The Control of Pollution Act 1974 was an initial UK attempt to bring together such medium-specific treatments into a coherent whole (Ball and Bell 1991) and was followed up in the 1980s with the consolidation of the UK regulatory agencies, a process which is continuing with the instigation of the unified Environment Agency in 1996. Since its third Environmental Action Programme (1982–6), the European Commission has pushed for environmental policy and its integration with other forms of policy (Klatte 1991).

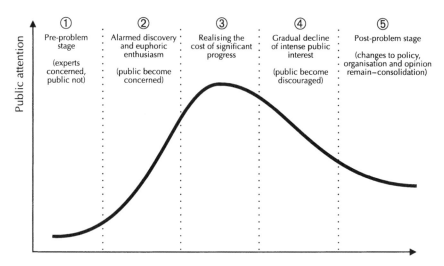

Figure 1.1   The 'issue-attention cycle'

Source: adapted from Downs, 1972

All these legislative moves both required and encouraged business, particularly manufacturing companies, to adapt their processes in line with environmental considerations. So, environmental attention became important because of its legal implications, but business did not yet emphasise the overall aim of 'greening' a company, beyond immediate regulatory compliance.

In the recession of the late 1970s and early 1980s, the environment featured less prominently on the political agenda. Buttel (1986) and Morrison *et al.* (1972) have suggested that environmentalism was sent into decline as its 'no-growth' message was more widely publicised, and that everyone converted (back) to growthism and supported business attempts to increase profits rather than protect the environment. There was undoubtedly a shift to economic issues in this period and this cyclic interest has been analysed prosaically by Downs (1972; interpreted graphically in Figure 1.1) as the 'issue-attention cycle'. Downs postulated that environmental concern, like other media issues, would rise in public prominence as people increasingly identified a problem and called for action. Concern would then reach a plateau with the dawning realisation of the costs of action, before descending again as people became discouraged by these costs. The argument in the 1970s was that environmental concern would follow this pattern, and therefore decline in public consciousness in the later part of the decade. This model downplays the role of external factors in suppressing concern by suggesting that it runs out of steam. For environmental concern, recession in the late 1970s and early 1980s served to focus public attention on economics not environmentalism, contributing to the dampening effects of the perceived costs of environmental protection.

However, we only have to look at the late 1980s to see that the environment was by no means destroyed as an issue in the mid-1970s, even if it was temporarily deprioritised. Downs asserted that environmental issues would evade the final decline in his model and he was right, as is evident from studies of media coverage of environmental issues (see Hansen 1993; Einsiedel and Coughlan 1993). But the style of environmentalism had again changed. There was less emphasis on societal questions of zero-growth and resource use, which spanned a multitude of industries and economies. Specific global environmental issues became foci for campaigning: the greenhouse effect, stratospheric ozone depletion and marine pollution replaced the earlier foci of pesticides and the spectre of fossil fuels running out. Individual actions were increasingly prioritised, replacing the more collective emphasis of organised environmentalism in the early 1970s. 'Public environmentalism' (Buttel and Larson 1980; Eden 1993a) developed where individual actions, as well as group protest, were given environmental importance. Green consumerism was promoted and 'green business' became a topic for public discussion, fostered by the attention paid to individual products rather than global economic restructuring.

Why was environmentalism different in the late 1980s? Like any sociopolitical idea, it had been adapted to fit new circumstances. The political and economic thinking of the time, dominated by the Thatcherite and Reaganite ideologies in the UK and USA, promoted market-led social change and individualism. This encouraged the development of green consumerism as an economic and individualistic way to influence wider (environmental) issues and militated against an all-out attack on the market ideology through criticisms of business activity. NGOs consequently diversified their campaigning into consumer-related actions, especially regarding product-specific global environmental issues, e.g. CFCs in aerosol sprays contributing to stratospheric ozone depletion. Products had been targeted before— Friends of the Earth campaigned against non-returnable glass bottles in the 1970s—but this time campaigns sought to change the habits of purchasers as well as of producers. However, such economic actions were soon complemented by the resurgence of more political and direct action for the environment, especially revolving around the anti-road campaigns in the UK in the early 1990s and the activities of Earth First! in the USA, so that environmental action was not restricted solely to an economically and individually oriented interpretation.

Whatever the impetus, by 1988 business had begun to evolve diverse tactics to address environmental issues: company policies, annual reporting, auditing, product labelling, advertising, partnerships and sponsorships all now carry environmental references for many large companies (although for small companies, environmental change is often precluded). From being subsumed under health and safety issues, environmental concerns were increasingly highlighted as a separate category by many companies during the 1980s, e.g. through special environmental publications.[2] Industry again attempted to portray itself as the solution to environmental problems (see Hoffman 1994 for an illustrative study of the oil and chemical industries). The International Chamber of Commerce, the Business Council for Sustainable Development and, in the UK, the Confederation of British Industry all became active in national and international debates. They sought to promote the interests of business in agenda setting and policy making, suggesting that 'industry may be much more pro-active on environmental issues in the 1990s' than it was in the late 1980s (Elkington 1990, p.10; Chapters 5 and 6).

Overall, we can see that the green banner was raised over many companies in the late 1980s as the rising tide of public environmental interest made the environment a topic for mainstream companies to build campaigns around. This occurred in sectors which were traditionally regarded as environmentally damaging, as well as the more specialised or 'concept' companies, such as the toiletries sector. The early 1990s saw the consolidation of these changes as the environment became incorporated within the wider set of social concerns acted on by business in the course of its activities. 'Green' was no longer a new colour by 1992, but it had become a commonly used adjective in boardrooms as well as on factory floors as industry sought more environmental input and influence.

Many things are changing as we move through the 1990s. International environmental negotiations are increasingly media-attractive, e.g. the United Nations Conference on Environment and Development (UNCED) in Rio de Janeiro in June 1992, so that the influence of the UN, the EC and other international bodies on business activities is likely to increase (Ball and Bell 1994). Environmental regulation is expanding and new means to enforce it are under debate, such as economic instruments developed in consultation with business: the landfill levy to be introduced in the UK in 1996, carbon tax proposals in the EU (Ikuwe and Skea 1994). These set policy targets for business to achieve, but build in latitude as to how business achieves those targets. Public interest in and expectations of environmental management are strengthening, as are schemes for its certification (Chapter 2). All these changes require us to look at how business is influencing and is influenced by the developments in the environmental debate.

WHY DOES BUSINESS WANT TO BE (SEEN TO BE) GREEN?

In the 1990s, then, business has paid attention to environmental issues in a more coherent and active way than previously. A variety of terms have been used to describe this development, including 'green business', 'business environmentalism' and 'corporate environmentalism' (e.g. Elkington 1994; Elkington and Dimmock 1991; Smart 1992; Karrh 1990). Business clearly wants to be (seen to be) green in the 1990s. There are three main reasons for this which I shall outline here by way of introduction, before pursuing them in more detail in later chapters.

Sales and savings

First, there is apparently profit to be made from 'green business'. Business writers stress this angle, arguing that business success comes at least partially from dealing with environmental problems (e.g. Taylor 1992). The tangible returns, the profits, are manifested in several ways. Direct sales may increase, although there is little widely available evidence at the moment that this is applicable to all product groups. Some companies have successfully marketed environmentally related products and increased their market share through so doing. For example, one company attributed a doubling of its market share to its use of the WWF's panda logo (Robins 1990) and Hoover reported that its washing machines, the first UK goods to receive the new European eco-label, had a 7% market share (*ENDS Report* 1993, 226, p.25; see Chapter 4). But in general, even where governments have contributed to the promotion of 'green' products, their sales have been low, e.g. energy-efficient light bulbs were targeted by the Energy Efficiency Office of the DOE in the UK because their sales in terms of units were only 0.2% in the UK in 1992 compared to 3.3% in Europe (*ENDS Report* 1995, 244, p.30). In fact, many companies have complained that market research has portrayed customers too optimistically, i.e. as more inclined to buy 'green' products than appears from sales figures; organic produce seems to have fared particularly badly in this respect (Eden 1992). Price premiums on products marketed with a green dimension may also provide tangible profits, but information on specific products is difficult to obtain. Safeway suggested in 1992 that it had reduced its profit margins on organic produce in order to help sluggish sales, but companies which have raised margins are less likely to publicise their activities (see Simms 1992). We can only suppose that premiums are generating profits but over only small sales volumes.

A second source of profit is where new markets develop, particularly for environmental services, e.g. consultancies and pollution abatement technologies. In 1992, the global market for environmental equipment and services was $200 billion, involved 60 000 businesses and employed 1.7 million people (*ENDS Report* 1992, 212, p.17). Taylor (1992, p.675) estimated that the market for environmental control technology in the USA and the EU alone was more than $73 billion per year and increasing by 6.5% per year, clearly a healthy market (and see Elkington *et al.* 1991). Indeed, statistics for the UK suggest that this sector outperformed mainstream business with a sales increase between 1985 and 1992 of 2.7% compared to the average for manufacturing industry of 0.8% (*ENDS Report* 1995, 248, p.3).

Thirdly, and probably more significantly in the short term, there are benefits to be gained from saving costs within the company, especially in production. 'Running environmentally cleaner often means running smarter and running cheaper' (Du Pont's CEO, in Smart 1992, p.191). This is part of 'green house-keeping' (Elkington *et al.* 1991), i.e. management that is efficient and practical in both environmental and economical terms. From a survey of 116 companies in the UK in 1994, Environmental Policy Consultants estimated that 40% of companies had made savings in production through purchasing environmental technologies (but 38% had not; *ENDS Report* 1994, 241, pp.4–5). For example, Safeway's energy conservation programme saved the superstore grocery chain about £1 million in the UK between 1988 and 1989 (Adams *et al.* 1990) and 3M's '3P' (Pollution Prevention Pays) saved them more than $530 million between 1975 and 1992 (3M in Smart 1992, p.14). Overall, cost saving is the most common rationale offered for environmental activity in the 'green business' conversion literature.

> There are opportunities for making money out of waste or emission controls, even where a firm is not traditionally in this business. There are even more opportunities for reducing costs, and hence enhancing profitability and competitiveness, by attention to detail. (the then CBI President, CBI booklet *Clean Up—It's Good Business* 1986, p.3)

Reputation

Less tangible than profits, a second reason to be green is enhanced reputation, including better image, good publicity and customer loyalty. These form part of an (indirect) investment in a company's long-term economic viability (Taylor 1992; Elkington *et al.* 1991) and have been linked to better recruitment of young staff who are perceived to be disdainful of companies with poor environmental names. However, reputation has far less immediate returns than sales and new markets. Hence, identifiable short-term costs are more readily accepted as investments in reputation where managers have a long-term view of the need for social viability. Environmentally related changes may also be consciously part of a parcel of measures which are being used strategically to (re)capture market share, especially when retailers wish to reposition themselves. This strategy was employed by Tesco when the grocery superstore retailer wanted to move 'up-market' in the early 1990s.

> The greening of Tesco is the result principally of research which showed that Tesco customers are becoming more sophisticated and critical. (Higham 1990a, p.17)

So, new customers were attracted to Tesco because of its new environmental image, but established customers were not lost, because they were becoming more environmentally aware themselves as part of wider public shifts. Indirect benefits therefore revolve around legitimation through public acceptance. The public can legitimate company 'greening' and express their approbation of changes most commonly through purchasing from that company. Therefore, the profit and legitimation reasons are tightly bound together.

Pre-emption

A third and final reason for business wanting to be green is in order to pre-empt public criticism and possibly also restrictive legislation. Poor or non-existent publicity is perceived to be detrimental, making the communication of positive activity necessary to avoid image problems. This defensive stand is made more important by the recent media and legal attention paid to 'bad' environmental news. Piesse (1992) has shown how such bad publicity can be more damaging indirectly through its effect on reputation than directly through the financial penalties imposed. For instance, she correlates the spill from the *Exxon Valdez* tanker off Alaska in 1989 with a fall in the international oil company's share price of $4–12, representing a total market decline in expectations of $5–15 billion. Yet the clean-up of the spill officially cost the company only $4 billion, even including the fine of $1.5 billion (Piesse 1992, p.50). Similarly, the leak of oil into the River Mersey from a Shell UK site in 1989 resulted in a fine of £1 million plus £1.5 million clean-up costs, but Piesse estimates that their shares lost £670 million (albeit temporarily). In a survey by Touche Ross (1990, p.9), over two-thirds of companies saw environmental issues as threats 'requiring defensive or corrective actions' rather than opportunities, due to the negative consequences of environmental 'accidents' on a 'good' company's reputation. It is in recognition of this that David Trippier (1990, p.11), when UK Minister of State for the Environment and Countryside, supported business 'greening':

> In this age of universal concern, it is in every business' own interests not only to be green, but to be seen to be green.

The more proactive and positive side of this pre-emption is where business involves itself in lobbying, putting its own interests forward clearly and forcefully in policy debates (Chapters 5 and 6) and where companies pressure their slower suppliers to pre-empt any contamination by association.

QUESTIONS OF COMPATIBILITY

> The linking of the marketplace to utopian ideals, to political and social freedom, to material well-being and to the realization of fantasy, represents the spectacle of liberation emanating from the bowels of domination and denial. (Ewen 1976, p.200)

There are clearly several dimensions to the business interest in environmental issues in the 1990s and several mutually reinforcing reasons for business activity on these fronts. But how far do these reflect a genuine compatibility between the interests of business and the interests of the environment? With the rising tide of public environmental interest in the 1980s came an expansion in publications attempting to establish this connection. Much was geared to consumers, encouraging them to adopt 'greener' behaviour, e.g. to buy recycled products, to recycle their own domestic waste and to write to their MP about road-building schemes. The consumer-pusher *par excellence* is Elkington and Hailes' *The Green Consumer Guide*, which in 1988 reached the non-fiction bestseller lists in the UK within four weeks of its publication (*The Economist* 1990, 8 September supplement). It was not alone: books and leaflets by Friends of the Earth, the Women's Environmental Network's Bernadette Vallely and *The Blue Peter Green Book* (Bronze *et al.* 1990), the latter aimed specifically at the audience of the popular children's television programme, also exhorted individuals to change, to become greener.

This consumer-oriented material was complemented by business publications. Industrial, retailing and service companies all produced their own material, to encourage consumer loyalty to their new 'green' credentials, to fend off attacks from environmental campaigners and to compete with their peers, especially the large companies who were increasingly adopting environmental dimensions in their marketing and reporting strategies (Chapter 4). A quantity of green business material has been produced, often building on early strands of social responsibility and self-regulation from the 1970s, but now explicitly focusing on the threats and opportunities of 'green business'.

Like the polemical environmentalist literature, much of this business literature had (and has) an evangelical aim (e.g. Welford 1994, p.120; indeed, Elkington 1994 begins with a quote from Pope John XXIII). Indeed, the USA Friends of the Earth President in 1992 questioned the validity of 'the gospel of corporate environmentalism' being spread by large companies. Associated with this evangelism was a plethora of 'opportunity-speak' in green business texts, which took a remarkably positive stance to the whole idea of 'green business', promoting it as an opportunity for business to extend its markets, to diversify, to legitimate its existence and activities, to retain autonomy in the face of threatened legislation and to profit by all these aspects. Examples of this pro-business stance are ubiquitous, but the better-known advocates are John Elkington and Tom Burke (exemplified in Elkington and Burke 1987; also Elkington 1990, 1994; Elkington *et al.* 1991; Burke and Hill 1990; Burke 1990). 'Today's apparent threats often flag up tomorrow's opportunities' is the overpowering theme (Elkington *et al.* 1991, p.223). But again, Elkington, though he was a good trend-spotter, was not the only one to ride this wave: texts by Schmidheiny and the Business Council for Sustainable Development (1993), Business and the Environment (Datschefski 1992) and Willums and Golüke for the ICC (1992) also reflect a pro-business stance and the fervent advocation of the environmental 'opportunity'. One main focus has been the potential of environmental reviews and consequent changes to save costs and therefore increase profits.

> By associating the solution to an actual or potential problem with the creation of new business opportunities, environmentally sustainable business development can be seen as a benefit rather than an additional cost. (Roberts 1995, p.94)

Commentators such as Simms (1992), Blaza (of the CBI, 1992) and Buck (of the Institute of Directors, 1992) have advocated business pro-environmental change because it is positively good for business (rather than because it is morally right).

Increasing awareness of the environment should be seen by business as an opportunity rather than a threat. In addition to the opening up of new areas of activity, environmental protection will provide benefits such as clean water and air as inputs to production, and cost savings from measures such as increased energy efficiency. (Buck, of the Institute of Directors, 1992, p.39)

More examples of 'opportunity-speak' abound, both in earlier exhortations of social responsibility (e.g. Stroup *et al.* 1987) and with respect to environmental responsibility. Moreover, much of this assumes that 'green business' is therefore not only possible and beneficial but that it *already exists* and that it is only through being (seen to be) green that companies will survive into the next century. This does more than offer the benefits noted above but argues that, without environmental considerations, business simply will not survive: the environment is now essential to business thinking.

The losers will be those companies which react too late, too slowly or not at all. For them, markets could be lost forever and their future viability irreparably damaged. (Macve and Carey, of the Institute of Chartered Accountants in England and Wales, 1992, p.5)

Generally, companies tend to labour the point that environmental and economic considerations can sit in easy balance, even that they complement one another in decision making and are compatible. They are attempting to integrate environmental ideas within their current operations and ideology but without serious overhauling of these—it is a conversion without pain.

The success of recent initiatives demonstrates forcefully that responsible environmental policies can be highly compatible with good business decisions and can enhance shareholder value. (Safeway Annual Report 1991, p.29)

Green business is therefore painted as a natural, inevitable and essential progression for companies. In a similar vein, Burke and Hill (1990) write of the 'interdependence' of economics and the environment and Elkington and Burke in *The Green Capitalists* (1987, p.169) state categorically that '[e]nvironmentally unsound activities are ultimately economically unsound'. The UK Conservative Government seemed to support this position of compatibility in its 1990 White Paper on the environment when it asserted that there was 'no contradiction in arguing both for economic growth and for environmental good sense' (Department of the Environment 1990, p.8). This viewpoint has fostered the expansion of environmental economics, especially since the production of the Pearce Report (Pearce *et al.* 1989) completed for the then Environment Secretary, Chris Patten, as input to the White Paper (and subsequently neglected, see Department of the Environment 1990, Annex). The Brundtland Report (WCED 1987) tackles similar themes (see Chapter 6).

These pro-business comments argue the case that business can easily and without conflict take up environmental issues because:

it is no longer the case that a company has to choose between economic efficiency and environmental quality; both are possible, and the achievement of one of these goals is likely to improve the prospect of achieving the other. (Roberts 1995, p.249)

In a similar vein, Elkington has recently (1994, p.90) described the potential for 'win–win–win' business strategies 'to simultaneously benefit the company, its customers, and the environment'. The general positive glow this sort of discussion gives to much writing is sometimes useful in attracting less convinced companies, sometimes amusing in its optimism and sometimes repetitive and ridden with clichés (e.g. Touche Ross 1990, p.21). At the extreme, this view produces a conceptualisation of business as the saviour of environmental problems, the solution to crises. John Elkington (1990 in CBI/ICC, p.17–18) suggests that 'business, rather than being seen as the enemy, is increasingly accepted as part of the solution' (and similarly Cooper in Hill 1992, p.1). The extreme of this view was put at a conference about environmental auditing with an audience of business people, by the then UK Minister of State for the Environment and Countryside, David Trippier:

> There is a lunatic view, held by some of the fundamentalist, 'dark' greens, that business is the enemy of the environment, and that the objective of making a profit is totally incompatible with the protection of our surroundings. I reject that view utterly. It is the product of well-meaning, but nevertheless totally unrealistic, minds which ramble down the road back to an imaginary Garden of Eden of 'airless', 'technologyless', 'industry-less', jobless individuals. That is not the real world and since it is the real world we are in the business of saving, we are entitled to dismiss that view as hopeless fantasy. (Trippier 1990, p.8)

Despite his fervour, this rather simplistic view has been disparaged by other commentators who point out the 'irreconcilable' (Higham 1990a, p.17) differences between environmentalist and business goals (e.g. Irvine 1989; Athanasiou 1996).

There has therefore been a considerable critical response to much of the literature cited above. One reviewer considered that '*The Green Capitalists* will appeal to those who want a concrete example of the way in which the system incorporates, makes harmless, and then makes a profit out of movements and ideas which seek to challenge it by selling the idea back to the activist in the form of one or more commodities' (McCulloch 1990, p.218). Such a critical response arises precisely because this material tends to argue for an inherent compatibility between environmental amelioration and business practices (e.g. Steger 1993, p.154). For environmental NGOs and other commentators, it becomes a nonsense to talk about 'green business', because it can only ever represent superficial or opportunistic change, predicated as it is on the capitalist ethos of using nature as a resource. Most business environmental change has been focused on operational change and environmental management systems (e.g. Hutchinson 1992; Ottman 1992). However, critics of this have sought to refocus their lenses onto the foundations and ideologies of business. Richard Welford has recently emerged as a pro-business commentator with a stronger environmental stance and he urges that a more fundamental change to business is necessary:

> The starting point must be to make the modern business enterprise challenge its very reason for being and to encourage, persuade, cajole or force businesses to take an ethical stance. (Welford 1994, p.18)

He advocates a more radical change to business than that offered by current environmental management systems because 'a piecemeal approach may satisfy the demands of the customers of some companies, it may well give companies a good feeling about some of the ethics involved in their activities' but it does not go very far towards truly sustainable business in environmental or equity terms (Welford 1993, p.30; and see Chapter 6).

Rob Gray (1990) has similarly taken a more fundamental stand to the changes needed. In Bebbington and Gray (1993), the authors argue that the current interpretation of environmental business is too restricted and that the dominance of accepted business ideologies has closed the debate on alternatives, such as total environmental accounting practices which appreciate environmental issues and fully discharge corporate environmental accountability. Only regulatory moves can revitalise such ideas because:

> if . . . there is a tension between social responsibility and ethics on the one hand and profit on the other, and the notion of profit is sacrosanct, then there can be no debate. (Bebbington and Gray 1993, p.1)

However, these writings are the vanguard of green business commentators, a state of which they are conscious. Most commentators are less radical in their claims for business and environment synergy and have concentrated on the positive cost–benefit calculations yielded by environmental reviews and accommodations. They have also proselytised about the compatibility between business and environmental concerns, as we have seen. It seems clear from even this cursory literature review that business–environment compatibility is being urged by business because of the advantages it sees therein, not on the basis of any ideological convergence. The compatibility is being constructed through argument and not being revealed through soul-searching: it is therefore a false compatibility, an issue we shall explore in subsequent chapters in detail but for which we can now suggest reasons.

IDEOLOGICAL DIVERGENCES

Claims that business and environmentalism are inherently compatible are dubious because there are some obvious obstacles to that compatibility. The sort of 'organised environmentalism' demonstrated by NGOs diverges greatly from that of

Table 1.2   Divergences in environmental thinking

	Business	Environmentalism
Primary divergences		
	Profit is priority	Environment is priority
	Continual growth	Low or no growth
	Expanding or changing consumption	Reducing and changing consumption
	Growth as end or objective	Growth as means to other ends
Secondary divergences		
	Short time frame	Long time frame
	Technical modification	Social, structural, lifestyle modification
	Pricing under present system	Restructuring pricing
	Information confidential	Open access to information

'corporate environmentalism' demonstrated by large companies in ideological (and practical) terms (Table 1.2).[3]

Fundamentally, commercial companies depend on the growth of the market and therefore a continuous increase in the sales of their products and services (Johnston 1989b), making market expansion a key goal for all businesses.[4] In the necessary relation between production and consumption, consumption must ever increase to soak up production output and this is achieved through generating new and varied needs to be fulfilled through purchasing (Galbraith 1972; Ewen 1976). Because modern manufacturing and retailing businesses rely on such (market) growth, the concept of limited growth (and especially a society based on a no-growth economy) is anathema, or at least alien, to the business mind. Also, the currently dominant business paradigm recognises no limitations on 'public goods', such as air and water, to which there is 'global open access' (Pearce *et al.* 1989) and for which there is no price to pay at the point of use.

Environmentalists take a very different line. One of the basic tenets of environmentalism is that the environment does not exist merely for people to exploit and use, but that there is a two-way relationship between nature and people and it is only by respecting this relationship that humankind will survive. If such a two-way relationship is the basis of reason, the excesses of unlimited growth and exploitation can only be seen as harmful. From the environmentalist point of view, therefore, increasing consumption, without incorporating higher goals of social and individual development or conservation of natural resources and sustainable production, must equally be anathema, so that '[c]onsumerism, of the kind represented by advertising and marketing, is largely incompatible with environmentalism' (Higham 1990b, p.17). In order to attack causes and not symptoms, many environmentalist writers believe that serious solutions to environmental degradation must involve social change, the de-emphasis of consumption, the support of public transport, new ways of planning urban and work environments, all to remove the need for many products and to reduce overall consumption without reducing the quality of life (see for example Porritt 1984; Kemp and Wall 1990; Green Party Manifesto 1987, pp.1–3). Some of these ideas are embedded in concepts of sustainable development and sustainability (Chapter 6), although there are different interpretations of these in circulation.

There is therefore considerable divergence between the objectives of environmentalist groups and those of industry—as expressed by a City stockbroker (quoted by John Vidal in *The Guardian* 12 January 1990, p.25):

> We don't find the want less, consume less and waste less approach of the Green Party particularly helpful because that obviously implies an economic recession, producing less [sic] manufactured goods and changing fundamentally the structure of society and manufacturing industry.

In contrast, environmentalists reverse this argument because they do not find the 'produce more, consume more' argument of business valid. Green consumerism has been criticised in this context because it de-emphasises *reducing* consumption in favour of *changing* consumption (and that only slightly—see Irvine 1989a,b). It fails to address the question of how much people consume because of its emphasis on what sorts of products they consume. Nor does it reduce consumption, which is environmentally damaging (Dobson 1990, p.141). Thus, environmentalist writers have described green consumerism *per se* as operating within the capitalist market-based system instead of changing this system (e.g. Dobson 1990, p.141; Gardner and Sheppard 1989, p.224; Pepper 1989/90).

There is a second source of divergence in the way businesses and environmentalists see means and ends. In a way, business goals are not truly 'goals' in the sense of being ends to strive for. Growth (of consumption and production) is a continual objective: there can never be a point reached at which growthists will cry enough (see Figure 1.2). Growth is an end in itself, whereas it could be seen as a means to some deeper or more distant objective. Environmentalists' objectives are often of this deeper nature, including social justice, democratic participation, freedom, equality and peace (for example, Spretnak and Capra 1985), which might use growth as a mechanism to foster sustainable development, but not as an objective (see Chapter 6).[5]

This fundamental conflict of vastly divergent goals gives rise to secondary, less fundamental but still serious conflicts between environmentalist commentators and business people. First, the two groups operate on different temporal frames (Birkin and Jørgensen 1994).

> One of the key difficulties which industrialists face in talking to environmentalists is that they operate on very different time-scales. Most businesses consider a two year time horizon a luxury, focusing instead on quarter-by-quarter results. (Elkington and Burke 1987, p.65)

Figure 1.2   The growth of economic growth

Environmentalists, meanwhile, are considering the long-term future of species and the environment perhaps centuries hence. To reconcile the two perspectives would require either a real sense of environmental crisis or fundamental changes in the character of the free market as it operates today, e.g. through the establishment of environmental legislation to force business restructuring.

A second conflict stems from the way that business frequently solves a problem through technical modification rather than a change of principles or structure. The classic case of this is the chemical substitution of CFCs in spray cans to attract green consumer interest to an 'ozone friendly' product. A step-by-step, technical approach is slow because it changes only particular elements of production incrementally, not systems of perception or behaviour. It is symptomatic of consumerist ideology that, when a product requires modification, the emphasis is on technological change or 'sticking-plaster environmentalism' (Irvine 1989a, p.23). Such a technical fix has been exposed as an insufficient, ineffective way to change attitudes and behaviour (e.g. Heberlein 1974) but is usually the one adopted by industry, such that 'the whole race for the Green car is another example of marketeers going for the cosmetic fix rather than facing up to the enormity of the problem' (Smith and Sambrook 1990, p.30).

So the goals advocated by the environmental movement of reducing consumption (e.g. Irvine 1989a,b) and putting the environment before profits are mostly alien to the practices of industry. Whilst companies may adopt green marketing strategies (Chapter 4) they ignore the real issues of environmentalism and the true ideal of green marketing—'the marketing of "less is more"' (Smith and Sambrook 1990, p.31). Hence, 'at some point, the irreconcilable philosophical divisions between environmentalism and conventional customer economics are bound to surface' (Higham 1990b, p.17). In the meantime, the conflicts of ends and means between commercial and environmentalist concerns are being papered over with green marketing. This is aided by bureaucracies able to master conflict, e.g. government departments, advertising and business watchdogs, which appear to be working for both sides of the argument but are implicitly working for the overall aim of maintaining the system itself (O'Riordan and Rayner 1991).

The third problem stemming from the fundamental divergence of goals is where conflict arises over the higher premiums on many green products, such as organic produce and biodegradable detergents. Such premiums are perceived by environmentalists as contributing to inequality because the 'green' consumer appears to be subsidising others who choose the cheaper, less 'green' version and therefore appears to be penalised financially for being environmentally 'good' (Irvine 1989b). This is because the full costs of production, use and disposal, e.g. pollution cleanup after production and landfill costs, are not incorporated into products but are regarded as externalities. There are then considerable difficulties in translating such externalities into market terms and therefore in identifying financial penalties and incentives for business change.[6]

This 'unreal' pricing, once conveyed to the consumer, is compounded by a class inequality where only the affluent consumers can afford to buy 'green' products because they are more expensive (Irvine 1989a). A related problem is that retailers are willing to raise 'green' product prices not because of increased production costs but because the demand will tolerate such prices. In other words, 'retailers are aiming at the committed green consumer, using market-skimming practices of higher pricing' (Simms 1992, p.39).

A fourth conflict is that environmentalists distrust business claims and information released about 'green' products and changes. Simms (1992, p.41) has described this suspicion as a backlash from consumers, environmental groups (such as Friends of the Earth) and the media in response to industry's opportunistic marketing. This is most evident in the environmentalists' denunciation of green advertising, e.g. the Green Con Awards for advertising that misinformed the public about products or used inappropriate claims or descriptions (see Chapter 4). The continued pressure from NGOs for greater provision and freedom of environmental information from businesses is indicative of this conflict.

So, there are several areas of divergence, together suggesting the impossibility of corporate environmentalism and organised environmentalism ever dovetailing in their approach to environmental activity. The locus of public environmentalism is

Table 1.3   Terminological positions in defining 'greenness'

Terms used to describe strong environmental positions	Terms used to describe weak environmental positions	Authors
Green	green	Dobson (1990)
ecocentric	anthropocentric	Eckersley (1992)
ecocentric	technocentric	O'Riordan (1981)
ecologist/ecological	environmentalist/environmental	Porritt (1984)
deep green	light/pale green	market research, see *Survey*, Winter 1989

probably somewhere between the other two ideologies, although more work needs to be done on this topic in detail. At least, it is clear that business environmental activity is regarded differently by different groups, meaning that the validity of the concept of 'green business' is presently being contested.

'GREEN BUSINESS' AS A CONTESTED CONCEPT

This gives us a problem if we wish to use the term 'green business' as shorthand for business environmental activity, because its meaning and applicability are ambiguous. For many commentators it represents a powerful contradiction in terms. We can appreciate this by briefly considering how 'green' has been defined in the environmentalist literature and seeing whether this applies to business at all.

The word 'green' is one of many that have been argued about at length and defined differently by authors theorising about environmentalism. The definitions of different authors particularly diverge on the question of how far 'green' implies a critique of modern society. Table 1.3 outlines some key terms from the literature in this area. For example, the 'anthropocentric' position of environmentalism is a human-oriented response to environmental problems which would include, for Eckersley (1992), resource conservation, because this takes a managerial and utilitarian attitude to the use of nature and prioritises human needs over environmental protection. However, an 'ecocentric' position would value species equally and not necessarily prioritise human needs over non-human ones. Similar general arguments are made by O'Riordan (1981) in distinguishing between 'ecocentric' and 'technocentric' positions (and see Rolston 1989). Distinctions have also been made between 'ecologist' and 'environmentalist', again to mark the level of change which is being advocated to economic, social and political systems (e.g. Porritt 1984). Although the details of these positions differ, they have in common with other commentators an overall interpretation of the level of 'greenness' as related to the support or critique of current social systems.

Dobson (1990) particularly discusses the use of 'Green' and 'green' to describe various orientations towards environmental protection in this light. He asserts that 'Green politics self-consciously confronts dominant paradigms' (p.5) whereas 'green politics presents no challenge at all to the late-twentieth-century consensus over the desirability of affluent, technological, service societies' (p.7). Ecologism, or 'Green' ideas, would therefore include radical actions such as restructuring the economic system, decentralising work practices, setting up a basic income system to replace state benefits, ensuring freedom of speech and proportional representation and other changes to the current welfare, economic, political and industrial systems (for policy examples, see Green Party writers such as Kemp and Wall 1990; Irvine and Ponton 1988; Porritt 1984; also Spretnak and Capra 1985).

In contrast, environmental NGOs which work within the system in order to gain greater consultative status and political acceptance are termed environmentalist or 'light green', green with a small 'g', e.g. the Council for the Protection of Rural England, the National Trust, the Royal Society for the Protection of Birds, the Sierra Club. These groups seek reform through environmental policy to ensure that industrial responsibility to the environment is fulfilled, but they do not seek to change the way industry is structured, e.g. to redirect attention from economic growth and increasing consumption to a no-growth economy and decentralised, de-automated production. Light green ideas echo terms such as 'technocentric' (O'Riordan 1981) because they put faith in science and technology as protectors of the environment. Tied to this technological orientation is a more managerial attitude to the environment, in the sense that natural resources are seen as available for use for human benefit.

Some NGOs are, inevitably, difficult to categorise. Friends of the Earth and Greenpeace might be classed as green by other groups who are more radical, e.g. Earth First!, but Green by business and more traditional groups, e.g. the National Trust. In reality, many environmental NGOs in the 1990s use both Green and green strategies, e.g. protests, petitions, research, publications, dialogue with business. Greenpeace is a good example because it maintains high profile confrontation on nuclear testing with business rapprochement on substitutes for ozone-depleting technologies (see Chapter 2) and with a hierarchical structure linking the grass-roots with the national and international offices.

From this brief description, we can see that a negligible amount of business activity fits Dobson's 'Green' politics because 'green business' still subscribes to the dominant paradigm and does not attempt to change the sociopolitical and economic systems which currently operate. Even with a small 'g', 'green business' does not seek to challenge the ideal of a modern, affluent, technology-driven and growth-oriented society. Rather, business and the environment are brought together in a positive way in the business literature solely for the benefit of business: 'no one [in industry] knows what green really means' but 'in many ways, green—regardless of how it is defined at any point in time—is good for business' (Ottman 1992, p.4). Business has appropriated 'green' in order to earn legitimacy and to ascribe a positive response to a multiplicity of changes brought about by external pressures (Chapter 2). Green business is therefore a blanket term, too widely appropriated for strict definition to be of practical or theoretical use.

In appropriating it, business has left behind the political and social bases of much environmentalist thinking. The concept has become diluted: '"green" is simply an adjective that has come to embody a wide range of opinions about the environment' (Robins 1990, p.147). In a similar vein, Welford (1994, p.18) uses 'green' to indicate an objective: 'what greening [of industry] ought to be' becomes the focus, not what it is now. Hence, 'green business' seems to be an inclusionary term not a discriminatory one. It is used in a relative sense, whereby companies who are behind in environmental terms see others as 'green' because they are environmentally proactive, e.g. in developing comprehensive environmental audits ahead of legislation (see Chapter 3). But to those outside business it is a blanket term for business activity which has lost any more specific meaning by its appropriation into business ideologies which diverge, as we have seen, from those of environmentalism.

THE SCOPE OF THIS BOOK

So far, I have reviewed the historical development of 'green business' and its claims to compatibility with environmentalism. In later chapters, I shall focus on particular elements of business activity on environmental issues, beginning with the forces for change and moving on to internal and external business developments and their implications for the environmental agenda.

Of course, the literature on 'green business' in general is still developing (partly due to the time lag of academic writing and publishing) so, although I am able to refer to many useful critiques of business, advertising, consumption and capitalism in general, detailed and in-depth critiques of green business are found infrequently (if at all). Hence in this book, a variety of academic and business strands will be intertwined to illustrate how green business has reiterated many old questions for society and has, in addition, raised new ones for environmental campaigners, regulators and consumers. This requires me to take an interdisciplinary approach to my central goal of critically examining 'green business'.

NOTES

1. For detailed discussions of strands in environmentalism, see Eckersley (1992); Dobson (1990); O'Riordan (1976); for USA conflicts specifically see Worster (1985).
2. Some companies still categorise environmental issues under health and safety considerations, thereby claiming them as part of 'the working environment' rather than part of a global external entity. Even so, they still tend to mention them more than they had done previously.
3. It is interesting to note in this connection that Elkington and Dimmock (1991, p.16) implicitly define

corporate environmentalism as a response to 'the need for business people to approach this area [of environmental performance] as *environmentalists*, rather than simply as professionals with environmental responsibilities'.

4. Although not the only goal—Galbraith (1972) suggested that autonomy, allowing a company to operate unfettered by government intervention, may have an even higher priority.

5. When we make this clear, it becomes difficult to see how the two perspectives might ever both be covered by the adjective 'green' (see later section).

6. This has been explored in the expanding literature on environmental economics. A reader unfamiliar with this literature might begin with Pearce *et al.* (1989); Pearce (1992); Jacobs (1990). Other studies include Helm (1991); Turner (1988); and for criticisms of environmental economics see also Eckersley (1993) and Athanasiou (1996).

2

Pressures on business for environmental change

The previous chapter introduced the issues raised by business interest in environmental issues, or 'corporate environmentalism', particularly its development, rationale and conflicts with organised environmentalism. In this chapter, I want to consider the pressures to which business is reacting when it takes environmental issues on board, before going on to consider the form of this reaction in later chapters. I will focus on four specific external sources of pressure: legislation; consumers; NGOs; other business sectors such as insurance, investors and suppliers. First, I should justify my choice of these four.

TYPES OF PRESSURE

Business often posits that its environmental activities are part of a reaction to outside pressures which force or encourage it to change, although some companies do claim that internal forces are involved as well (see Chapter 4). However, we begin to get into difficult territory once we try to assess which factors are most significant in the 1990s because surveys addressing this question have provided different answers.

By way of illustration, consider a quantitative survey of 116 companies in the UK in 1994 by Environmental Policy Consultants, 90% of companies questioned cited UK legislation as the driving force for environmental change, 62% cited corporate environmental policies, 57% cited European legislation, 40% cited cost savings, 16% cited consumers and 9% cited environmental NGOs.[1] Compare this with the results from a survey of over 500 company directors for *Director* magazine for the Institute of Directors in 1990 (Nash 1990; Table 2.1). The survey pre-chose a list of pressures on directors to develop an environmental policy, namely: employees; customers; suppliers; public opinion; family (the last two being identified as the most influential). The survey did not ask about legislative pressures, which are highly important in bringing about business change and were the most significant by far in the previous survey, or explicitly about environmental NGOs

Table 2.1 Responses to the question 'Have you found yourself under pressure from the following groups to develop an environmental policy?'

Pressure source	All sectors	Manufacturing	Non-manufacturing
Public opinion	50%	52%	50%
Family	48%	39%	53%
Employees	25%	21%	28%
Customers	21%	26%	18%
Suppliers	7%	9%	5%

Source: adapted from Nash (1990, where n>500, multiple responses possible, all figures rounded to nearest percentage)

Table 2.2 Sources of pressure on companies for environmental change

Drivers of environmental performance, most significant first	Influences on company behaviour, most significant first
Government legislation	Local authorities
Corporate social responsibility	Local community
EC legislation	National pressure groups
Customers	Media
Commercial pressures	Local pressure groups
Public opinion	
Conscience	

Source: adapted from PA Consulting Group survey for CBI, in CBI 1990 (responses ranked in order of significance)

(although they might be included in 'public opinion'). This view of pressures is therefore a very restricted one. From the table alone, we might suppose that the external pressures of business reputation, indicated by the opinion of customers and the more general public, were perhaps the strongest issues in encouraging business environmental change and therefore that legitimation is the primary objective of the changes.

A contemporary survey (Table 2.2) by PA Consulting Group for the CBI in the UK questioned 250 companies (large companies, judging by the comment that they represent 40% of UK GDP by turnover) about the pressures they felt. Here, the sources of pressure are divided between influences on behaviour and on performance, the latter dominated by legislation and 'internal' ethics, the former by different 'publics' surrounding the company (compare the survey of German companies in Steger 1993). Again, UK and European legislation features highly in the pressures on environmental action.

Roome (1992) also distinguished sources of pressure in categorising business vulnerability. He assumed that legislation is driven by public perception in the immediate term and only more slowly by scientific developments. In his scheme therefore, as in Nash's, legislation is not a principal or explicit part of the pressure but is a consequence of other pressures. Roome's classification is twofold, using the scientific significance of environmental issues against their public perception to divide company responses into four categories. For issues that are not very significant on either scale, business action is discretionary, i.e. unpressured, and consequently, we might add, unlikely. Roome, like Nash, suggests that companies respond more quickly to change in public perception than to change in the scientific understanding of environmental issues, which only elicits a slow response, but in either case he still characterises company strategies as reactive.

Like most surveys, therefore, the three discussed produce results determined by their own assumptions. By investigating transsectoral samples, the surveys do not distinguish the pressures specific to companies with particular environmental implications. For many retailers, legislation has been less forceful than public opinion in prompting them to review their environmental activities. For manufacturing, and especially small and medium-sized enterprises (SMEs), legislation is far greater in influence, and certain sectors, e.g. chemicals, oil and refining and energy generation, see themselves as particularly 'at the sharp end' in this respect. Furthermore, different external factors clearly are mutually influential: legislation is often prompted by public support, NGOs and national and international politics; NGOs depend on public opinion and financial support; suppliers and other business customers are influenced by external factors as well as then turning their influence on other companies within their production chain.

In justifying my choice of four sources of influence in this chapter, therefore, I can only call on them as umbrella terms for a range of factors which are not necessarily mutually exclusive nor uniformly felt by business. *Legislation* is clearly important because it is often cited alongside business changes, e.g. CFC reduction is linked to the Montreal Protocol controls (see later), but it depends on the political climate. *Consumers* link to public opinion, which is important for company legitimation as noted above, although again this differs significantly by sector. *NGOs* have been identified by some of the surveys above but they are rarely directly cited by companies as sources of pressure, perhaps to downplay their role as critics to be countered. They are, however, influential in directing and affecting legislation and consumers through campaigns, although their successes cannot be shown conclusively because of the other factors in the social context. *Other businesses* are noted in the surveys and are becoming increasingly important as environmental investment and insurance begin to appreciate the implications of environmental legislation, e.g. in the oil industry. Often this fourth source reflects the first because, for instance, legislative and political developments around liability for environmental damage are worrying investors and insurers, who consequently put pressure on companies to disclose their impacts more clearly.

So, there are connections between the four sources. They might even be seen as sources of pressure for each other as well as for business, but they are all clearly important to the direction and degree of business activity on environmental issues and considering them in detail will allow us to go further than the survey snapshots above. I shall first deal with legislative (and political) pressures, which I consider to be the most important across all business sectors because they require compliance by all, not merely the most responsive, largest or highest profile companies. They also usually require remedial action, not merely remedial publicity as may be the case where customer or NGO pressure must be countered. I shall then deal with consumer demand, NGOs and other pressures within industry. In each case, I shall explore how pressures on business developed as well as their implications for its environmental activity.

THE LEGISLATIVE AND POLITICAL CONTEXT

One of the most fundamental pressures for business environmental change is that of legislation and regulation. It is not, however, immune to business influence— business is customarily consulted during the drafting of legislation and also lobbies to change, reframe and delay environmental regulation, often successfully (see Chapters 5 and 6). Touche Ross (1990, pp.20–21) call legislation a 'driving force' for company policy, but then note that half (32) the companies they surveyed said that they were unaware of any legislation likely to affect them.

Political influences on the regulatory climate

First of all, we should note that legislative pressures for environmental change are themselves influenced by the political climate. In the early period of 'green business' activity in the UK, the rising fortunes of the green parties of Europe were particularly influential, as was the international scientific consensus over ozone depletion in 1987–8. It is worth exploring these changes briefly to put the legislative and regulatory pressures in some context.

Increasing public political involvement and radicalisation of society in the 1960s (Chapter 1) heralded the evolution of the political arm of the environmental movement in the 1970s. However, it initially saw little success, with the UK's green party (then People) achieving only 1% of the vote in 1979. Gradually through the 1980s 'green' voting became more widespread, particularly in those European countries whose political systems incorporated proportional representation. Greens gained seats in the national parliaments of Switzerland in 1979, Belgium in 1981, Luxembourg in 1984, Austria in 1986, Finland in 1987, Italy in 1987 and Sweden in 1988 (Porritt and Winner 1988; Rüdig and Lowe 1986). In 1984, Greens were elected to the European Parliament from West Germany (7 in all) and from the Netherlands and Belgium (McCormick 1989, p.142), and Green parties in many countries have also gained local seats, e.g. in Belgium, Switzerland and Luxembourg (Spretnak and Capra 1985).

However, it is the German Greens, *Die Grünen*, who have had the strongest European presence since they were founded in 1979. In 1983, 27 Greens took their seats in the Bundestag, the first new party to achieve over 5% representation (and therefore seats) in 30 years. But there have been problems with the stability of the Green constituency. The German Greens failed to re-enter the national parliament in December 1990 and there followed a period of conflict between those advocating deep changes to the party to revitalise its presence (Fundis) and those advocating working with the present system to gain incremental changes (Realos).[2] However, *Die Grünen* pulled themselves up, despite the death of their influential speaker, Petra Kelly, and re-entered the Bundestag in June 1994 with 10% of the vote, 3000 local councillors and a potential for coalition with the SPD, having resolved the damaging Fundi–Realo conflicts. Their story reflects the recurring strength of the environmental agenda in Germany, which has in turn led to innovative legislation regarding, for example, packaging responsibilities for industry (see Chapter 5).

In the UK, success has been more ephemeral. The high point of Green voting came in the European Parliament elections in 1989, when the UK Green Party achieved 15% of the cast vote, making it 'the most successful Green Party in Europe' (*Guardian* 1990, 28 September, p.34). However, this was an isolated incident, probably related far more to protest votes against the Conservative Party and the perception that the election was divorced from 'real' national politics, than to deep and lasting support. Since that election, the Green vote in the UK seems to have disintegrated, with the party losing all its 253 deposits in the 1992 General Election with only 1.3% of the vote (*Greenline* 1992, No.97) and Party membership slipping below 5000 in 1994. There are still some 300 Green local and parish councillors and Cynog Davis was elected as an MP in 1992 with joint support from the Welsh National Party (Plaid Cymru) and the Green Party, but the likelihood of any real Green power in the House of Commons is extremely small.

Hence, the UK Green Party looks for its influence more in terms of making other parties assess environmental issues more carefully and as a pressure group within the general environmental umbrella. Moreover, it certainly seems that its 1989 success shook the mainstream parties temporarily, coming as it did on top of media interest and the increasing attention being paid to ozone depletion by atmospheric science and international negotiations. Mainstream parties had already responded in a minor way to the earlier wave of environmental concern, e.g. in 1977 both the Liberal Party and the Conservative Party set up ecology groups (Flynn and Lowe 1992), but the spate of environmental controversies prompted Margaret Thatcher, after denying that the environment was a real issue (McCormick 1991), to make a speech to the Royal Society in 1988 claiming environmental issues as Conservative ground:

> It is we Conservatives who are not merely Friends of the Earth. We are its guardians and trustees for generations to come. (Flynn and Lowe 1992, p.27)

This speech has been emphasised in the literature (with hindsight rather more than it was at the time) for its attempt to gain the political high ground in the face of the increasing scientific consensus over climate change and ozone depletion (see later). This was reflected in the publication of a government white paper entitled *This Common Inheritance: Britain's Environmental Strategy* (Department of the Environment 1990) and the passing of the Environmental Protection Act 1990, under the patronage of Chris Patten, the supposedly 'green' Secretary of State for the Environment at the time.

Since then, recession has overtaken environmental issues on the political agenda, as have wrangles over European unity and international issues in Eastern Europe and elsewhere. This might suggest that 'green business' is now publicly forgotten, but the seeds sown in 1989–90 were the product of both 20 years of environmental concern, albeit at lower levels, and anticipation of rising environmental concern at the end of the recession. The process of environmental regulation slowed, but it did not cease. In the UK, it became consolidated and incorporated more specifically some of the issues raised in the 1990 white paper, leading most significantly to:

1. a national waste strategy and associated recycling plans from local authorities, the Producer Responsibility Initiative (see Chapter 5) and other more publicly oriented initiatives;
2. Integrated Pollution Control programme specifications through Guidance Notes and now the integration of HMIP (Her Majesty's Inspectorate of Pollution), NRA (National Rivers Authority) and some local authorities' duties in the new Environment Agency.

However, regulation and enforcement depend strongly on the government's disposition towards environmental reforms. In the UK, the Conservative Government is ideologically opposed to strong regulation of business, preferring market-based mechanisms and arguing that the latter offer more efficient, cheaper, faster and more flexible ways of causing business to deal with issues such as the environment (Flynn and Lowe 1992). Further, a broad deregulation initiative begun in 1993 was widely supported by industry, e.g. the Institute of Directors and the Chemical Industries Association (CIA), the latter arguing that environmental regulation was particularly hurting the chemical businesses. Environmentalists feared that environmental deregulation would be 'a smokescreen to dismantle the framework of environmental regulation' (*ENDS Report* 1993, 217, p.18) even before the Environmental Protection Act 1990 was fully bedded in. In fact, the Deregulation and Contracting Out Act 1994 did not deal intensively with environmental legislation, but it did 'make a sweeping and wholly unsubstantiated assertion' (*ENDS Report* 1995, 248, p.28) that environmental legislation was a burden on business. Moreover, it required the new Environment Agency to consider how to minimise that burden in terms of reducing the costs to business of environmental regulation. (Whilst this is hardly its major function and will inevitably hamper its environmental effects, it does, however, fit precisely into the divergences noted in Chapter 1).

Legislative and regulatory influences

Environmental legislation in the UK has always had a distinctly discretionary nature (Ball and Bell 1991), meaning that its enforcement has built-in flexibility for business in terms of how it achieves policy targets. The prosecution of Shell UK by the National Rivers Authority (NRA) following pollution of the River Mersey in 1989 was seen as a chance for the newly established agency to show its teeth and its new 'arm's-length' relationship with business, and resulted in a £1 million fine for Shell UK on top of clean-up costs (see Carruthers 1993). A specific aim of the Environmental Protection Act 1990 was to address the 'cosy relationship between inspectors and inspected' (*ENDS Report* 1990, 186, July, p.11) through distancing HMIP, and now the Environment Agency, from the companies. Until this is successfully implemented, it is likely that close links will continue in the UK between regulators and business, despite business's complaints that the government considers channels for communication between business and government as less important than business does (Grant 1993).

Other countries have developed stricter regulatory regimes, e.g. Denmark, Germany and the USA. The US Environmental Protection Agency (EPA) can enforce regulation more effectively and enhance the impact of legislative moves. The so-called Superfund legislation is perhaps the most contentious environmental regulation, established after the Love Canal incident in New York (1979) as the Comprehensive Environmental Response, Compensation and Liability Act 1980 (CERCLA). Under this, the state can force those parties which it identifies as responsible to clean up hazardous waste sites. Liable parties include the current site owner, the site owner at the time of contamination (even if waste was disposed of correctly under legal provisions pertaining at the time of disposal), the waste producer, the waste disposer on site, the waste transporter and the waste broker. Liability is 'strict', i.e. allocated regardless of fault or negligence, and 'joint and several', i.e. it can be based on ability to pay not on the contribution to contamination, quite a fundamental change. It is also retroactive in that it applies current technological capability to the clean-up of past contamination (Robins 1990, p.56). Superfund built a trust fund from taxes on oil and chemical industries in order to pay for site clean-ups; this was expanded under the Superfund Amendments and Reauthorization Act 1986 (SARA), which provided $9 billion for emergency clean-ups. A Superfund National Priorities List is kept of sites needing emergency clean-up and the parties which are potentially responsible are pursued by the EPA. The largest single cost recovered by 1990 was $66 million for a site in California (Robins 1990, p.55).

CERCLA and SARA have changed the business climate for environmental issues in the USA to a far more confrontational and investment-conscious one. SARA's Title III is particularly important for environmental reporting. Established in the aftermath of the leak of gas at a Union Carbide chemical plant in Bhopal, India, it requires that companies report all hazardous or toxic releases, including those which the company was allowed under legal permits, a process which has often exposed problems to ill-informed company managements as well as to the public. Altogether, this legislation represents the strictest environmental laws on businesses. However, liability is often contested, so that:

> thousands of companies are spending billions of dollars not on cleaning up contaminated sites but on lawsuits against the EPA, Justice Department, states, counties, municipalities and, most important, their insurance brokers . . . many argue that the only parties cleaning up under Superfund are the lawyers. (Robins 1990, p.56)

Such conflicts over attributing liability have meant that 'Superfund has apparently so far generated more legal action and lawyers' fees than cleaned up toxic waste sites' (Sethi 1990, p.4).

The Superfund legislation enacted in the 1980s in the USA has repeatedly threatened to surface in European or UK liability legislation. In 1989, the EC proposed a Directive on civil liability for damage caused by waste, which would have strict as well as joint and several liability (i.e. individual and collective) but would not operate retroactively, therefore ultimate liability would not be clearly defined (Robins 1990; Simmons and Cowell 1993). An EC green paper in May 1993 suggested harmonising environmental liability across the Community on the basis of 'strict liability'. A later report from the UK's main business association, the CBI, responded that '[l]iability for remedying environmental damage may well become the key environmental challenge facing business in the 1990s' (*Financial Times* 1993, 3 November, p.15) but did not accept that business should be responsible for paying for the clean-up of contaminated land. In particular, the CBI was horrified by the retroactive powers of the USA's Superfund legislation because it means that pollution legal in one year may become a liability in later years under higher regulatory standards—this makes the regulatory context of business decisions considerably more uncertain over long time periods. The tracing of responsible parties over time is also problematic (*Financial Times* 1993, 3 November, p.15; *ENDS Report* 1993, 225, pp.19–22). The CBI favoured the idea of contamination costs being raised through grants from a mixture of public and private sectors, effectively relocating at least part of the responsibility for clean-up to the state. The UK government statement in response to the EC paper rejected joint and several liability and compulsory insurance against compensation claims, and generally supported business and the CBI's concerns about costs and possible market distortions. In the light of the US experience, the EC is unlikely to be able to overcome strong business and domestic government opposition to a coherent position on liability. Business has generally demanded that there are no changes to the liability status quo, in order to guarantee both stability and certainty for its environmental decisions (*ENDS Report* 1993, 225, p.19; ACBE 1994).

As well as national legislation, a prime source of environmental regulation in recent years has been the European Commission (EC). Although the environment was not specifically covered in the original Treaty of Rome which established the European Economic Community, it was introduced into the recent five-year plan

(1989–94) and into the 1987 Single European Act (SEA) which laid the foundations for a single market in Europe.

Several points should be noted about the external pressures from the EC on business environmental change. First, EC legislation is rarely binding on member states—only where Regulations are issued does the European specification become national law. More commonly, when Directives are issued by the EC, member states must take the responsibility for translating these into national law and enforcing them, and obviously this depends on the national government's inclination and ideology. In fact, by 1991 no Directive had been implemented by all member states within the time period allowed (Klatte 1991). This throws up our second problem: European legislation commonly depends on agreement, which is often hard to come by in Europe. The Danish no-vote on Maastricht has been cited by some as indicating the Danish worry that their more stringent environmental standards would be undermined by the input of less environmentally concerned member states. Thirdly, the EC is attempting to integrate environmental policies with other policies, e.g. those on employment, economics and development (Klatte 1991). This would imply that the environmental issues are to be assessed alongside economic issues, rather than to act as a constraint on them, something which should encourage business to consider legislation more positively. Fourthly, European business lobbying is developing and, in the course of encouraging policy integration, the Commission is establishing channels through which business can more effectively communicate with legislators, in the hope of achieving less confrontational processes. This will presumably enhance the influence of business on EC regulation and therefore may soften the pressure in the long term.

Because of the different policy regimes and administrations within the Community, business does not suffer uniform environmental regulation pressures. Klatte (1991) has classed member states into four groups of countries in decreasing order of environmental activity:

- Group 1: Denmark, Germany and the Netherlands (ahead of EC standards);
- Group 2: Belgium, France, Italy and Luxembourg (indifferently complying);
- Group 3: UK and Ireland (reticent in implementing European standards);
- Group 4: Greece, Spain and Portugal (more recent members, therefore lagging behind).

The amount of environmental regulation on business roughly parallels these trends. It is likely that two of the new members of the European Union, Sweden and Austria, will soon approach categorisation in Group 1 due to their commitment to environmental regulation.

Neither is the regulatory influence uniform across business sectors. From the earliest environmental regulations in the nineteenth century, legislation has focused on industrial processes, so that manufacturing has more commonly been at the 'sharp end' of environmental regulation and hence more active in attempting to influence it than retailing and commerce (Chapters 5 and 6). In the UK, production regulation was initially dealt with by separate bodies for emissions to land, to air and to water. From the late 1980s, the situation has been more coordinated: HMIP was responsible for industrial emissions to air, water and land under Integrated Pollution Control (IPC) and for radioactive waste. The National Rivers Authority was responsible for water pollution more generally, in inland, underground and coastal waters (excluding those emissions covered by IPC) and local authorities were responsible for noise and nuisance control, waste disposal (mainly to landfill) and small-scale sources of air pollution not dealt with under IPC (Ball and Bell 1994). From April 1996, the Environment Agency has integrated many of these functions.

The customary discretion employed by regulatory agencies is beginning to break down in the aftermath of the Environmental Protection Act (EPA) 1990, although not as quickly or as thoroughly as environmentalists might have hoped (see Chapter 5). Because the EPA was 'enabling' legislation, it left the fine details to be worked out once it was implemented. Consequently, HMIP found itself dependent on industrial input to its technically based Guidance Notes, especially regarding the definitions of best practicable environmental options (BPEOs) and best available techniques not exceeding excessive cost (BATNEECs), because industry had the specialist research and development and practical experience to comment in detail on draft notes. Indeed, industry has argued that the very vagueness of BPEO in the Act and subsequent discussions is the cause of the poor take-up of the concept in business practices (*ENDS Report* 1994, 236, p.8).

In many cases, these kinds of pollution regulations depend on technical specifications and are generally seen as constraints on businesses, requiring a compliance-oriented response. Indeed, it has been suggested that a much lower proportion of Danish companies have written company environmental policies than UK companies because the former have to comply with much more voluminous and stringent environmental legislation, leaving no incentive for voluntary initiatives (Touche Ross 1990). Where there is less legislation, there is more opportunity to pre-empt the possibility and a more proactive response is facilitated. Also, larger companies and more influential sectors have more influence through pre-emptive action and lobbying (see the case study of Du Pont in Doyle 1992).

Most of the regulation discussed here relates to production, i.e. factory or in-store processes, and has the heaviest impacts on manufacturing sectors. However, product regulation may have greater impacts on non-manufacturing sectors, such as regulations dealing with eco-labelling (Chapter 4) and recycling (Chapter 5). Again, regulatory pressures have differential impacts according to business size, sector and stance.

The example of ozone depletion

An example of regulation and its effects will illustrate these issues: stratospheric ozone depletion. This became big international news in 1985–6, when studies from

Antarctica exposed a marked depletion in the concentration of stratospheric ozone over the pole in the Antarctic spring (Mazur and Lee 1993). The Montreal Protocol was drafted in 1987 to respond to these concerns by setting national reduction targets for the compounds behind the depletion. In order to achieve international agreement, it allocated grace periods and funds for technology transfer to signatory countries in the south. The Protocol initially set targets of 50% reduction in national consumption of CFCs by the end of the twentieth century. The continuing scientific evidence, public opinion and the energies of the media and environmental pressure groups combined to push governments and industries towards a more rapid phase-out of a wider range of chemicals under reviews to the Protocol, especially in London in 1990. Together with the rapid development of competition in the non-CFC domestic spray can market particularly, by 1990 90% of UK aerosols were CFC-free and other areas of CFC use in business became foci for action, such as refrigeration and blown-foam packaging which used types of CFCs (e.g. Kemp 1993).

The Montreal Protocol is therefore often cited as a successful international environmental regulation—unfortunately one of the few—because of its widespread support, rapid implementation and acceleration of controls. But this seeming consensus and rapid change belie a number of processes, pre-emptive actions and reactions which deserve closer scrutiny. First, it had a longer history than is often realised. There were initial bans on CFCs in spray cans in the 1970s in countries such as the USA, Sweden and Canada, after early worries about the effects of CFCs on stratospheric ozone (Elkington and Burke 1987, p.136). Because of a subsequent decline in public attention and therefore political pressure, these bans never resulted in wider business development of alternatives to CFCs apart from in spray cans (Dudek *et al.* 1990). In the mid 1980s, when scientific evidence pointed to CFCs as culprits of substantial stratospheric ozone depletion (Mazur and Lee 1993; Ross 1991), the lack of conclusive proof was used to delay action. However, when the very newsworthy, colourful and visible maps of Antarctic ozone depletion were publicised in 1986, public opinion and support for environmental campaigns to stop CFC use gathered force and governmental attention intensified in response (Purvis 1994). The international policy climate was similarly affected, producing agreement on the Protocol on CFCs and halons in 1987. The rules of the Protocol incorporated frequent reviews of its targets because of the uncertainty and rapid evolution of the related scientific and industrial information.

Secondly, the CFC-producing companies[3] attempted to soften the impact of the Montreal Protocol during its drafting by lobbying national governments and attending international discussions. Industry representatives, principally of the chemicals and aerosols sectors which were directly affected, were present during UNEP meetings to draft the Protocol prior to the Diplomatic Conference at which it was to be discussed by national government representatives. Since then, industry representatives have continued to meet with officials to influence the Protocol's implementation, e.g. the UK aerosol industry association met with EC officials regarding its implementation in Europe (BAMA Annual Report 1987).

Following the initial responses to the Protocol, in 1988 the scientific consensus was secured by the measurement of large ozone reductions in Antarctica, and the continuing media attention was undoubtedly important in securing acceleration of the phase-out schedule under reviews in 1990 and 1992 (Benedick 1991). Now industry could see that the legislative die was cast and the Protocol had obtained considerable international political and public support, forever labelling CFCs as damaging in the public mind. At this point (and not before, see Doyle 1992; Karrh 1990, p.72) manufacturers and retailers reoriented their activities to the positive so that they might respond to the threat to their corporate reputations. The producers felt threatened by 'a massive consumer backlash, orchestrated by pressure groups such as Friends of the Earth and fuelled by a stream of new scientific findings against CFC' (Dewhurst 1990, p.67). The response was both research and development into substitutes and new public relations developments, e.g. codes of practice and similar positive programmes of corporate policy change (Peattie and Ratnayaka 1992; see Chapter 5). These often built on research and development into substitutes for CFCs in spray cans, begun in the late 1970s. However, this time political and public support for change was sustained and, having accepted the inevitability of a halt to CFC production, companies invested in technological modification in order to respond to this demand.

One example is Du Pont, one of the world's biggest CFC manufacturers (*Financial Times* 1994, February 23, p.18; Karrh 1990, p.72) and the employer of the discoverer of CFCs in 1928 (Doyle 1992, p.87). The company announced in 1988 (the year after the Montreal Protocol was signed) that it would stop producing CFCs completely (Du Pont CEO in Smart 1992, p.186; Doyle 1992, p.88). It also publicised its subsequent investment in research into and development of CFC substitutes at a cost of £300 million (*Financial Times* 1994, February 23, p.18).

> Our unilateral decision to phase out the manufacture of [CFCs and halons] as soon as possible—but by the end of the 1990s at the latest—went beyond the Montreal Protocol and has set the pace for an accelerated phaseout by other manufacturers. (Karrh, director of Du Pont, 1990, p.72; and see Du Pont CEO in Smart 1992, p.186)

Critics have seen this move as purely a reaction to oncoming legislation:

> only when the cause [of preventing legislation] is clearly lost, will Du Pont pre-empt anticipated legislation by announcing over the nation's airwaves its decision to clean up. (Doyle 1992, p.87)

Indeed, in 1974 Du Pont had said that it would stop CFC production if science could prove that CFCs were damaging to health. This was less a serious promise than a defiant rejection of contemporary criticisms. Nevertheless, Du Pont was reminded of this statement by US senators in a letter three weeks before its 1988 declaration to cease CFC and halon manufacture (Benedick 1991, pp.111–12).

So, by 1988, the political and public pressures had become unavoidable. Retailers in many countries called on manufacturers to phase out CFCs in aerosol products, did so in their own-label brands and directed consumers towards CFC-free aerosols. The turnaround by industry from lack of interest and delaying tactics to enthusiastic promotion and research was swift. In fact, it has been argued that the ease of chemical substitution of CFCs, e.g. by HCFCs and methyl chloroform, and the research and development already conducted in the 1970s allowed companies to accelerate the phase-out schedule originally negotiated under the Montreal Protocol, although this has been continually denied by industry representatives. Also, as the big companies took the lead, the others were forced to follow their competitors so that the impact of CFC phase-out was felt throughout the production chain. It seems that by 1990 the UK government, for one, was encouraged to call for faster phase-out on the advice of its major producers, who were not about to see their investments in substitutes being rendered useless by the continued survival of CFC-containing competitors. Ease of substitution and the competitive imperative, stimulated by scientific and public consensus and further urged by international politics, fed back into politics and encouraged acceleration of the Protocol's phase-out targets. Businesses then pushed to expand the CFC-free market through product labelling and other promotional activities—sales of HFCs were expanding by 1994 as the Montreal Protocol restrictions took hold (*Financial Times* 1994, February 23, p.18).

Moreover, the technological input changed as the scientific consensus on which gases were ozone depletors changed. At first, HCFCs were used as CFC-substitutes as they are similar in structure and thus could perform similar functions. However, they are also similar in their capacity to destroy ozone molecules, leading to their rapid identification as ozone depletors and calls for them to be controlled under the Protocol and to be discouraged as CFC-substitutes (*Financial Times* 1994, February 23, p.18; Benedick 1991, p.175). HFCs, which contain no chlorine and thus do not destroy ozone, were adopted as CFC-substitutes in refrigeration, especially HFC134a in industrial chilling and air conditioning (rather than in domestic fridges, which are only 5% of the total refrigeration sector). But the recent identification of HFCs as contributing to global warming because of their potential as greenhouse gases introduces uncertainty over their industrial future. This has led to businesses lobbying to maintain the viability of their investments in HFCs, especially with regard to controls on greenhouse gases under the UN Convention on Climate Change (*ENDS Report* 1995, 242, p.10).

However, the more unlike CFCs substitutes become, the more difficult is the substitution process because it requires greater adaptation of equipment, e.g. air conditioners in cars. Kemp (1993, pp.94–5) suggests that a third of the current market for CFCs will disappear as non-CFC technologies emerge (some being cheaper and easier than CFCs, e.g. soap and water as a substitute cleaning mix), another third will be replaced by 'CFC-likes' (HCFCs and HFCs) and a final third will be made redundant by the efficient use and conservation of 'CFC-likes'. So the overall technological change is probably restricted to one-third of current usage, the remainder representing merely modifications to current operations (see Purvis *et al.* 1995, p.19).

CFCs in particular provide an interesting case of the two-way flow of influence, from science to public and politics, to business and back again, and of the complications of reaction and proaction that they demonstrate. After initial reluctance and inertia, businesses became strong supporters of further acceleration as they invested in alternatives. ICI, the major UK chemicals manufacturer, claimed to be 'at the forefront of international efforts to solve the [ozone] problem' as part of its positive and 'open' public relations campaign (Dewhurst 1990, p.70). Both business associations and environmental pressure groups claim to have (almost single-handedly) brought about CFC phase-out. Robins (1990, p.146) suggests that the UK Friends of the Earth's consumer campaign which effectively stigmatised spray cans which contained CFCs 'played a significant role in forcing the aerosol industry to agree to phase out CFC use by the end of 1989'. But others claimed the credit for business change. The British Aerosols Manufacturers Association (BAMA) said that in February 1988 it had:

> recommended that its members should stop using CFCs in their products as rapidly as possible and no later than the end of 1989. As a result of the rapid response by aerosol manufacturers *to that recommendation*, virtually all consumer aerosols produced in the UK, do not use CFCs. (BAMA leaflet, *Aerosols: your questions answered*, n.d., p.5, emphasis added)

What actually happened was that eight companies which were members of BAMA chose to phase out CFCs voluntarily and BAMA supported them *after* this was announced. Previously, the UK aerosol industry had been criticised for failing to accept the new political and scientific consensus about ozone depletion (Benedick 1991, pp.103–4). Its claims are also undermined by its adherence to previous reductions: BAMA has continually referred to a decrease of 30% in CFC usage in Europe between 1976 and 1981 and attributed this solely to positive change in the aerosol sector, without acknowledging the national policy pressures which focused entirely on the aerosol sector during this period. Hence, aerosol production had been pressured to change in the past, but BAMA was using this as an example of independent and laudable activity to pre-empt further pressures on its particular sector (for example BAMA Annual Report 1988, p.6).

As with many publications by trade associations and companies, BAMA publicly claims that the promise by eight of its members was due to internal company culture, apparently nothing to do with either the Montreal Protocol signed the previous year, or consumer or NGO pressure. We can safely say, however, that the changes were not entirely internal, nor entirely driven by such trade associations, but were the outcome of a two-way process of influence which is often not explicitly acknowledged. As Doyle (1992, p.90) notes for one large company, 'Du Pont's environmental image is deliberately constructed to suggest that industry is capable of keeping its house in order without government interference', and yet,

'Du Pont would not be cleaning up today, or setting goals for waste minimization and emissions reductions, if it were not for government action'. The different sources of pressure clearly feed one another: NGOs feed public opinion, which can feed political action and regulation and also be influenced by them in turn. Businesses are neither purely victims nor dictators in this process but seek their influence where they may.

CONSUMER DEMAND

The second external pressure on which I shall focus is that of consumer demand, which reflects public opinion specifically through economic channels of purchasing. This is often cited as the primary pressure, the 'main driving force' (Blaza 1992, p.33) for business change on environmental issues. That is debatable: often public opinion (and therefore consumer opinion) feeds into legislative development (as Roome 1992 assumes in his analysis of company change) which is more *directly* influential. For example, in the survey by the CBI and Environmental Policy Consultants at the beginning of this chapter, consumer demand lagged behind legislation as a driver of change. Also, in a small-scale company survey in 1990, Touche Ross (1990, p.15) reported that only half the companies had altered or were planning to alter their products in order to meet consumer demand. It is therefore important to assess how consumer demand has developed in relation to environmental issues and its potential for business coercion.

Green consumerism

As mentioned in Chapter 1, the recent wave of environmental concern in Europe showed far stronger business and consumer connections than had previously been the case. Public environmentalism was increasingly expressed through financial channels, e.g. donations to and passive membership of NGOs and, as we focus on here, purchases. Since the late 1980s, green consumerism has often been cited as a key motive for 'green business'. However, it is a rather ambiguous concept, partly because of the employment of the single term 'green' to describe a multitude of environmental activities (see Chapter 1) and partly because of the dual meaning of 'consumerism'. 'Consumerism' has been used to describe the social trend towards a 'culture of consumption', where wants are constructed and advertised in such a way as to encourage their satisfaction through (individualistic) purchase (e.g. Sinclair 1987, p.65; Ewen 1976; see Chapter 4). The concepts of the 'leisure industry' and the 'heritage industry' are in this mould—previously 'free' activities are now packaged for private consumption. In this sense, 'consumerism' is a pejorative term, encapsulating materialism and a culture dependent on individualism and mass production.

In a nutshell, consumerism equates more possessions with greater happiness. You are what you own, and the more you own, the happier you will be. (Irvine 1989a, p.15)

In its alternative incarnation, 'consumerism' describes the fight for consumer rights, for clear and trustworthy information from business and for discerning purchasing, embodied by consumer pressure groups such as the Consumers Association in the UK. Along these lines, *The Ethical Consumer* magazine (1989, 4, p.13) hoped that green consumerism had 'introduced a huge audience to the idea that consumers are to some extent responsible for the consequences of their buying decisions'. This second meaning represents criticism of the first, especially the emphasis on increasing consumption and production without reference to environmental effects (Sinclair 1987, p.65).

So, 'green consumerism' means different things to different groups. Business has focused on a positive reading of its second meaning, i.e. pressure from discerning consumers looking for environmentally friendly products and better environmental information. Criticisms of business from politically motivated, collectively organised and activist consumer pressure groups and boycott campaigns, who are seen as 'unrealistic' in their demands (see later section), are thereby separated from positive and individualistic consumer pressures. For example, Adams *et al.* (1990, p.3) describe green consumerism as 'the exercise of consumer choice which expresses a preference for less environmentally harmful goods and services'. This emphasises the importance of *choice*: green consumerism is characterised as individualistic and decentralised decision making on the basis of identified self-interest (Smith 1990, p.30). This distinguishes it from consumer boycotts, which have been going on for some time (see Smith 1990), because it is positive: preference rather than refusal is being stressed. It is also essentially an individual action, whereas boycotts depend on more collective communication and action and a voicing of protest. It is possible that the continued emphasis on economic protest through consumerism will decrease the adoption of political protest in modern society (e.g. Hirschman 1970). Therefore, although at base an economic act, green consumerism has significant social ramifications (Vogel 1975 quoted in Smith 1990, p.182).

In some countries, green consumer activity has a long history: (West) Germany has identified environmentally less harmful products by the Blue Angel mark since 1978. In contrast, green consumerism only became important in the UK in 1988, when the media were covering the issues of ozone and tropical rainforests with zeal and *The Green Consumer Guide* (Elkington and Hailes 1988) became a bestseller. Interest was boosted by the unprecendented and short-lived UK Green vote in the European elections (Johnston 1989a), but by 1990 the peak had been reached, recession was taking hold and environmental issues began to wander down the policy agenda. From the peak of newspaper coverage in 1989–90 (Simmons 1993), green consumerism entered a phase of subdued but continual media interest and a more sceptical public attitude to its promise. The credibility of environmentally friendly products was widely attacked on both environmental and pricing fronts (see Chapter 1) and public disillusionment set in. A lower degree of interest was sustained through the early 1990s, but it is quite difficult to measure—few statistics exist that specifically measure the 'green market'. Many environmental products are produced by companies which also produce non-environmentally marketed products, e.g. detergents groups, supermarket own-labels. In this case, statistics on company share of the market tell us little about these environmental niches. Neither do they really reflect consumer demand for environmental products, beset as such statistics are by other variables, such as availability (especially with respect to organic produce), pricing premiums and trust problems. Equally, consumers buying one environmentally friendly product may simultaneously buy another that is not, making the application of the term even more problematic.

Measuring the green market

Despite this, during the green consumerism peak of 1989–90, market research tended to focus on measuring the overall proportion of green consumers in the purchasing population in the UK and therefore suggesting the overall size of a 'green market'. This was again a positive move, to encourage business to undertake change in order to exploit this pool of consumers. However, the different market research companies produced different extrapolations of the level of public participation in green consumerism from their sample surveys, according to how 'green' was defined (see Table 2.3). The most quoted estimate in 1990 (and since) was MORI's, which suggested that 42% of adults had participated in green consumer behaviour, producing a green market size extrapolation of 18 million adults, quite an encouragement to business. However, since this was defined on the basis of people who had purchased *at least one* product because of environmental considerations, it was rather broad and tended to include rather than exclude consumers. It also therefore included those who had tried such products for novelty or for a short period, rather than those who had changed their lifestyle to commit to such products. It is even more difficult to find a definitive measure of ethical consumers who not only positively choose 'green' products but also boycott damaging products within a systematic buying pattern. By assessing these and similar figures (e.g. *Survey*, Winter 1989), I would suggest that perhaps 10 to 20% of adults buy *some* sort of 'green' products regularly and might constitute a tentative 'green market'. Even so, because of the dynamism of the situation and the broadness of such definitions, it is unhelpful to take such estimates as absolute, especially as consumer behaviour varies considerably between different products and issues. For example, a UK Department of the Environment poll in 1989 noted that 64% of its sample had bought ozone-friendly aerosols but only 25% had bought recycled paper and only 9% phosphate-free washing powder (*ENDS Report* 1994, 239, pp.23–4). This makes estimates of green consumerism very difficult: it often seems that estimates of green consumer behaviour monitor public opinion towards the environment more clearly than they do economic behaviour.

Table 2.3   Estimates and definitions of green consumers in the UK

% adults sampled that fit definition	Definition	Company and year
60%	make a positive effort to buy environmentally friendly products, termed 'dark greens' (c.41%) + buy environmentally friendly products if they see them, termed 'pale greens' (c.21%)	Mintel 1994
49%	actively seek green products	Mintel 1991
42%	had chosen one product over another on the basis of environmental performance at least once	MORI 1990 (and again in 1994)
30%	actively seek out environmentally friendly products rather than just occasionally preferring them, termed 'green thinkers'[4]	Diagnostics Market Research 1990
28%	'bought products because they were environmentally friendly on a regular basis'[5]	NOP for Department of the Environment 1993

Sources: Elkington (1989); *The Times* (1989, 30 June, p.5); *International Journal of Retail and Distribution Management* (1992, p.v); Burnside (1990); *Survey* (1989, Winter); *ENDS Report* (1994, 232, pp.18–20; 1994, 239, pp.23–4)

Since the 1989–90 peak, therefore, green consumer demand has declined somewhat, but it has not disappeared. Retailers and other commentators have suggested that environmental products are following the same path as 'healthy eating' products in the UK in the mid 1980s. In this case, there was an initial rush of interest and demand, which prompted an expansion in product ranges and manufacturer and retailer activity, followed by consumer scepticism over the claims and the higher prices of healthy eating products. Later, as their profile declined, these products became accepted as the 'norm' and developed a solid demand, if at a smaller volume than initial interest might have predicted (similar to Downs 1972 cited in Chapter 1). This represents a two-way process between consumers and business, where business capitalises on consumer interest but this capitalisation is then attacked by rising scepticism, which eventually balances out in a small but established market. The final stage is perhaps developing now for green products.

The rationale for green consumerism has been explored little in the literature. There is a general assumption that it is similar to involvement with environmental groups, which has been connected to people's value systems, occupations, education and social status (see studies by Van Liere and Dunlap 1981; Samdah and

Robertson 1989; Hopper and Nielsen 1990; Cotgrove 1982). However, green consumerism is more individualistic than participation in organised environmentalism through groups. It constitutes a form of public environmentalism where public sympathies for groups are shown in a diffuse and mainly financial way. More important for green consumers than value systems and beliefs in the morality of environmentally sound activity are the impacts of their purchasing decisions (Eden 1993a). The basis of the phenomenon is a utilitarian one, and its future growth may depend on the impacts of less environmentally damaging products being convincingly communicated rather than solely on public environmental concern.

Consumer sovereignty

So far I have outlined the rise and extent of green consumerism, but we now need to turn to how business views this trend in order to be able to assess its contribution to business environmental change. Business frequently refers to consumer demand as an external pressure by using the language of consumer sovereignty—that consumers dictate their wishes to business which then acts on this demand. We need to look at this argument to understand how far consumer demand represents an external pressure for business change and how far it is a two-way process of influence, i.e. the prevalence of reaction or proaction in this case.

Consumer sovereignty rests on the assumption of a one-way flow of instruction from the consumer to retailers, manufacturers and government bodies. Thus, the consumer is sovereign in the sense that the sum of consumer purchases serves as a 'social edict' (Galbraith 1972, p.166) to be acted on without question or choice by their suppliers. Galbraith (1972, p.218) termed this assumption the 'accepted sequence' with its belief that 'the individual is the ultimate source of power in the economic system'. This is implicit in writings by Simms (1992), Heelas and Szerszynski (1991) and advocates of green consumerism as a force for change such as Adams *et al.* (1990), Burke (1990) and, most famously, Elkington and Hailes' *The Green Consumer Guide*:

> We should not underestimate the power that each of us has today to start changing the world by operating as Green Consumers. Money talks—and there is growing evidence that major companies are waking up to the commercial potential of Europe's green markets. (Elkington and Hailes 1988, p.323)

However, there are problems with the assumption of such consumer power in that it is limited by an individual's situation. Where information or choice is constrained, sovereignty cannot be fulfilled because people do not have the information on which to make choices (Adams *et al.* 1990; Heelas and Szerszynski 1991; Smith 1990, p.295). As business and other institutions play a large part in providing both information and choice, their actions curtail consumer sovereignty, causing a two-way flow of instruction. The assumption of consumer sovereignty also ignores ways in which business both influences consumer preferences and directs purchases through product choice. Consumer sovereignty has therefore been widely criticised (e.g. Galbraith 1972; Schnaiberg 1980) for neglecting the power of business to influence and constrain markets for its own profit and autonomy. Galbraith drew attention to the flow of instruction from business to consumer, terming it 'demand management'. He described a revised sequence to replace consumer sovereignty in which business and other institutions promote social beliefs and individual goals which favour themselves, e.g. through promoting the continued consumption of manufactured goods. Ewen implicated the advertising industry as the particular agent of this promotion of consumer culture:

> Consumerism, the mass participation in the values of the mass-industrial market, thus emerged in the 1920s not as a smooth progression from earlier and less developed patterns of consumption, but rather as an aggressive device of corporate survival. (Ewen 1976, p.54)

This reversal of consumer sovereignty has been termed producer sovereignty by Smith (1990, p.34) and is reflected in Gorz's critique of the development of capitalism in the last century when it became apparent that:

> *consumption would have to be in the service of production.* Production would no longer have the function of satisfying existing needs in the most efficient way possible; on the contrary, it was needs, which would increasingly have the function of enabling production to keep growing. (Gorz 1988, p.114, emphasis in original)

It seems clear that consumer sovereignty is too simplistic a notion, but it continues to be ubiquitous in the business literature. Moreover, the difficulties of teasing out consumer and business influence in a two-way flow of instruction are illustrated by Adams, who invokes both consumer sovereignty and business demand management in the same sentence:

> The constant factor . . . is individual demand. This demand, though in part created by them, is the tune to which our major companies dance. (Adams 1992, p.107)

So, some commentators in the field of ethical and green purchasing seem to embrace consumer sovereignty but with partial recognition of the role of demand management by business.

A further criticism is that consumer sovereignty is used by capitalist institutions and actors to legitimate their actions (Harte *et al.* 1990; Smith 1990). Power is supposedly transferred from the business to the consumer, therefore characterising business as purely reactive and relieving it of any responsibility, rendering it supposedly amoral (Gorz 1988, p.112). Any changes are portrayed as originating in demand and therefore the power of business is downplayed. Advertising and marketing implicitly or explicitly use consumer sovereignty to justify actions which can be criticised on moral grounds (Smith 1990; and see Chapter 4).[6]

It is difficult to state conclusively how we should view the contribution of consumer demand to the greening of business. I have shown above how consumer sovereignty is invoked to argue that business follows consumer trends and therefore that external pressures dictate how far business becomes 'green'. Under this argument, the consumer, not business, is responsible for its level of environmental change, because only what green consumers buy can be sustained by business. This rarely acknowledges the role of 'demand management'. The availability, pricing (see Chapter 1 on market-skimming practices) and nature of green products are dependent on business, such that consumers can only purchase within the confines of these conditions. More importantly, business can legitimate its lack of change by arguing that consumer demand does not support the widespread penetration of green products, and therefore place the responsibility for lack of business change onto consumers rather than business.

But business also professes internal changes, initiated by company culture rather than by consumers (see Chapter 3). This is a reversal of the consumer sovereignty argument above, and would seem to suggest internal pressures for change. However, because consumers must approve business changes through purchases and enhanced reputation, internal changes cannot exist without external approbation. Thus, neither consumer sovereignty nor producer sovereignty is alone adequate to understand the sources of business environmental change. Consumer pressure is undoubtedly important, but it operates in a two-way process of legitimation for business change which is often underemphasised.

> Green consumerism, in short, has shown itself to be a double-edged sword—as much a new source of pressure on business to clean up its act as a novel marketing opportunity. (*ENDS Report* 1990, 180, p.2)

ENVIRONMENTAL NGOS

Voluntary environmental activism and the activities of non-governmental organisations (NGOs) or pressure groups also influence business environmental change, albeit less severely than legislation, and also consumer demand and public opinion on environmental issues. Much academic work has studied the development of environmental NGOs (e.g. Lowe and Goyder 1983; McCormick 1989; Yearley 1991), their philosophies (e.g. Dobson 1990; Eckersley 1992; Goodin 1992) and the people involved in them (e.g. Manzo and Weinstein 1987; Cotgrove 1982). For my purposes, I shall look at how environmental pressure groups influence green business rather than solely the development of those groups.

Recently, companies and NGOs have developed partnerships (Forrester 1990; Bendall and Sullivan 1996) and businesses have expressed their willingness to enter into dialogue with their critics in NGOs. However, this may represent merely 'the new environmental etiquette of the 1990s' (Doyle 1992, p.90) rather than the convergence of the two groups. Environmental NGOs are still, in the main, outsiders to business, from which point they have claimed success in changing business activities. Friends of the Earth claims to have influenced the chemical industry over CFC use and the horticultural industry over peat products from protected sites of special scientific interest (SSSIs) in the UK. *ENDS Report* (1990, 184, p.15) noted the success achieved by environmentalists in pressuring industries to reduce CFC production ahead of the Montreal Protocol requirements, especially in spray can and foam-blowing technologies. Indeed, it refers to change in ICI's UK manufacturing operations as 'cultural change in which the voice of the marketing professionals is being heard more strongly while that of chemical engineers has diminished'. It also notes that Friends of the Earth's call for consumers to boycott CFC-containing aerosols prompted manufacturers to advertise their products as 'ozone friendly' in 1988. Although this pressure is unlikely to have achieved anything substantial without scientific consensus and the regulatory backing of the Montreal Protocol, it did serve as a very public expression of concern which capitalised on and reinforced the other forms (see earlier in this chapter).

The environmental groups influence business both directly, e.g. through lobbying offices, stores and corporate HQs (see *Earth Matters* 1991, 13, p.7) and indirectly through influencing public demand, the political agenda and media sources. Much of group expenditure goes on campaigning resources, e.g. research and publication of findings from research as well as leafleting, posters and education. Friends of the Earth accounts for 1994 show that over £3 million was spent on campaigning and information services (*Annual Review* 1994/5) and this is comparable to the smaller business associations (NGOs themselves), e.g. The Paper Federation in the UK, but much less than the budgets of other business organisations, e.g. the major cross-sectoral association in the UK, the CBI, and many of its multinational members. Through such information, the ideas of the élite or minority of activists within such groups are fed down to what Lowe and Goyder (1983) term the 'attentive public'. These people may be passive members of groups (only donating, not taking non-financial action) or may not be members but feel sympathetic to the aims of such groups. This filtering down of information (see deHaven-Smith 1988) allows the organisational élites to target specific issues for individual action by the membership and the public, as well as more organised group demonstrations and activities. The results of this process can often affect business where the activists target products or companies for members' (financial) action. For example, Friends of the Earth linked rainforest hardwood logging to deforestation and damage to indigenous peoples in its 'Mahogany is Murder' campaign. In the UK, this publicity led to a campaign for sustainable forestry products and the boycotting of tropical hardwoods focused on DIY stores. In turn, the retailing sector passed responsibility down the chain to its suppliers, compounding the NGO influence with an intra-business pressure (e.g. B&Q 1995). Similarly, Friends of the Earth mounted a sustained campaign against Fisons for extracting peat for garden compost products from UK SSSIs. When Fisons announced that it would be selling its horticulture business, the Friends of the Earth journal promised no let-up:

> Friends of the Earth can take much of the credit for making peat extraction such a thorn in Fisons' corporate body over the past two years . . . Prospective buyers [of the horticulture business] should take note that Friends of the Earth will campaign against any company seeking to profit from the wrecking of our remaining lowland peatbog habitats. (*Earth Matters* 1992, 15, p.5)

This illustrates the two-way nature of the influence between organised environmentalism and public environmentalism and their combined effect on business. As well as group activists identifying issues for public consumption and action, the public and passive members of groups supply monetary support to the activists and allow them to claim representativeness as a voice of public opinion on environmental issues. In Lowe and Goyder's (1983) study of environmental pressure groups, the activists and élites of the groups, e.g. group secretaries, treasurers and chairs, saw the passive membership of groups as an instrumental resource, so that the 90% 'lump' are directed and used as support by the estimated 10% activist membership.

At present, perhaps 5 to 8% of adults in the UK are dues-paying members of environmental organisations (McCormick 1991, p.152). Estimates vary because many members, perhaps as many as 60% (Lowe and Goyder 1983, p.36), belong to more than one group simultaneously, so that additive totals of group memberships count a large proportion more than once. Regardless of the actual proportion, it is clear that environmental activity is still seen as a rather eccentric, and certainly minority, interest. If participation was to broaden to include more people from different sectors of the population, the impact of public opinion and NGOs on business would intensify. Some commentators feel that this has already happened, that environmental activity has become more mainstream and conventionally accepted (e.g. Burke 1990) and therefore a more central concern for business. This seems to be rather optimistic in the light of comments made by media and business people who continue to stereotype environmental activists as nostalgic technology-haters and utopia-seekers (David Trippier as quoted in Chapter 1). In return, Irvine (1989a,b) and Pepper (1989/90) have produced vitriolic criticisms of green business as 'sticking-plaster environmentalism' which addresses only the symptoms of environmental problems in the weakest way, and only where this supports the industrial aim of indefinitely expanding production and consumption. The main divergences between environmentalism and business views have already been outlined in Chapter 1 and we can now add a divergence in terms of support. Environmental NGOs have smaller, less reliable and more diffuse sources of income than do business associations with their large, dependable corporate subscriptions. This has given rise to a fundamental concern about the respective influence of business and environmental NGOs, especially whether environmental

NGOs can match the media, public and governmental support achieved by the better resourced and more mainstream position of business.

This issue is compounded by the nature of public environmentalism in the 1990s. Purchasing is a basic stimulus behind most business in modern society, and the purchasing of more by its customers is the mechanism by which business survives. Moreover, the public may feel more comfortable with expressing support financially, rather than through political protest, because it is depersonalised and easily chosen or discarded (see Hirschman 1970). Also, business draws much of its lobbying authority from economic arguments, whereas NGOs focus on political analyses, moral choices and values. The contrast is particularly great between business NGOs and those environmental NGOs that have taken a more radical and theatrical line in campaigning, e.g. Friends of the Earth, Greenpeace.[7] But in a modern society which prioritises economic arguments, particularly in the current neo-Conservative political climate, economic arguments tend to influence policy more easily and authoritatively. It is in the interests of industry to deflect attention from the cumulatively high levels of material and energy consumption towards arguments about which products—all of which must have some sort of environ-mental impact—reduce or increase environmental damage. In other words, business concentrates on solutions rooted in technology and economics rather than the restructuring of the system which has created these problems. Some environ-mentalists have characterised such a deflection as a seizure of initiative by indus-trialists, according the latter greater influence in directing the environmental debate.

> Right now the environmental movement has not only lost the initiative, it has allowed the debate over environmental action to be re-framed by government and industry. (Rose 1990, p.1)

Moreover, the external pressure from environmental NGOs on business may be declining as business lobbies strengthen and influence the environmental debate more strongly, and as economic issues are prioritised in the prolonged 1990s' recession.

Whilst business usually reacts to pressure groups, it is increasingly interested in building bridges (Forrester 1990) in order to anticipate and accommodate criticism from external sources. With increasing proactivity in the legislative arena, the relative positions of business and environmental NGOs are likely to fluctuate further (Chapters 5 and 6). In reciprocal fashion, environmental NGOs have developed ways of working positively with business in private meetings, although their public confrontations may remain antagonistic. For example, Greenpeace produces a newsletter, *Greenpeace Business*, which began as a joint venture with Friends of the Earth to influence the electricity privatisation in the UK in 1989 through persuading corporate investors of the environmental implications. Now the group has taken a 'two-edged sword' approach to relations with business, incorporating 'positive solutions' alongside 'traditional confrontation' (*Greenpeace Business* 1993, 16, p.1). This has included supporting the major superstore retailer

Safeway when it agreed to join the boycott campaign against Norwegian produce (instigated when Norway resumed commercial whaling in contravention of the International Whaling Commission's moratorium). The newsletter even asked Greenpeace members to 'shop at Safeway'. Further, Greenpeace has been involved in the development of a 'greenfreeze' fridge and was 'largely responsible for re-introducing hydrocarbons as a viable alternative to CFCs and HFCs' (*Financial Times* 1994, February 23, p.18). In this case, Greenpeace criticised industry for continuing to support HFC134a as the successor to CFCs because it was another ozone depletor. It developed the new fridge design using hydrocarbons (butane, propane and cyclopropane), independent research and production being provided by a former East German company, DKK Scharfenstein (now Foron). The first models were delivered in Germany in March 1993, complete with a Blue Angel endorsement (see Chapter 4), and by 1994 they had taken 5% of German market share. In 1992, seven large (West) German companies, e.g. Bosch-Siemens, publicly criticised the product, but by February 1993 they had developed their own versions, suggesting that Greenpeace's industrial foray had had a positive and effective influence (Verheul and Vergragt 1994). This again implies that Greenpeace is turning away from confrontation and public debate as ways to influence industry and moving to 'positive persuasion' of manufacturers by playing their own game. By late 1994, more than half of the European production (and 15% of global production) of fridges used some form of 'greenfreeze' technology, rather than CFCs or HFCs, according to Greenpeace's statistics. The technology is not 'new' so cannot be patented—one reason, Greenpeace argues, for companies rejecting it at first (until pressurised). Greenpeace stresses the consequent industrial opposition it encountered in the name of 'big business':

> The greenfreeze story shows that genuine change is possible and that even powerful industrial strangleholds can be broken. (Greenpeace *The Greenfreeze Story* leaflet, n.d.)[8]

As well as such positive campaigning, Greenpeace has continued its con-frontational approach to business activities. In contrast to its support for Safeway, it has toured the sites of a competing UK grocery superstore chain, Tesco, with a lorry containing new non-HFC-cooled supermarket freezers and carrying the caption 'Tesco's freezers wreck the planet' (*ENDS Report* 1995, 240, p.29). Tesco brands this 'misinformation', based on false claims about the level and availability of such 'greenfreeze' technology.

We can see from this section that environmental NGOs have some influence on business directly via boycotts on products and companies and indirectly through stimulating and supporting public opinion and through lobbying for environmental regulation (in which they have probably been less successful in the 1990s, see Chapters 5 and 6). But they are continually reforming their strategies both in opposition to and partnership with business in the 1990s, again complicating the categorisation of reaction and proaction.

INVESTMENT, INSURANCE AND OTHER SECTORAL PRESSURES

> Finance is that last great nut for the environmental movement to crack . . . Environmentalists are beginning to realize that great power that individual and institutional investors could exert to promote higher standards of corporate environmental performance. (Robins 1990, p.162)

As well as the pressures on industry from legislation, consumers and NGOs, there is some pressure to assure financial investors and insurers that company 'greening' is genuine and profitable. Increasingly, company policy and action are being assessed in the share market and have an influence on company viability. Therefore the need to improve this representation is an additional pressure on business to respond to environmental concerns. It is likely, however, that the impact of negative publicity, e.g. through accidents, is far greater than that of positive publicity through environmental action (Piesse 1992; see Chapter 1) so that the avoidance of this, as well as the dissemination of positive information, becomes very important in legitimation. Environmental performance and reporting will influence companies' financial attractiveness and legitimate their activities not only to consumers but also to insurers and investors.

The ethical investment movement began in the USA, developing out of social and environmental concern in the 1960s (Robins 1990, p.162). This included environmental concerns over pollution records but also employment practices, links with oppressive political regimes, animal testing and connections to armaments manufacture and supply. In the 1980s, ethical investment entered the financial mainstream in the USA, with most brokerage houses having in-house ethical investment advisers. Its value rose from $40 billion in 1984 to $450 billion in 1989, based on screening out companies with poor environmental performances (Robins 1990).

Specifically environmental funds emerged from the ethical investment movement in the 1980s. For example, a Stewardship Fund for environmentally friendly investment was launched by Friends Provident as early as 1984 in the UK (Miller 1992). Rather than solely screening out 'bad' companies, this invested in companies which met 10 criteria of positive environmental awareness and action. Initially valued at £0.5 million, it had grown to £150 million by the end of 1989. Overall, ethical investments and environmental unit trusts burgeoned in the UK in the late 1980s (Harte and Owen 1991) and by 1993 there were over 20 companies offering more than 30 ethical funds in the UK (EIRIS 1993). More than £400 million is invested in the funds and between July 1991 and February 1993 they grew by over 15%, whilst unit trusts as a whole grew by only 3% (EIRIS 1993). Other European countries have developed similar funds, although in Germany legal obstacles prevent the screening out of companies on non-financial criteria (Robins 1990). Studies in the USA have suggested that stocks tend to underperform for companies with old and inefficient technology or management which is not responsive to environmental costs (see Tanega 1993). However, the present lack of a ratings system on environmental grounds underplays the importance of such considerations in financial performance.

The pressure for business environmental change from these kinds of investment funds comes from requirements not only for better environmental performance but also for better environmental disclosure on that performance. Disclosure becomes more necessary and comprehensive and encourages change so that the information disclosed is favourable. 'Companies will often find that they are on a green treadmill, with expectations constantly increasing' about environmental reporting (Robins 1990, p.165). In a sense, this investment pressure depends on other investment companies but also on their customers, so that we might see it as another form of consumer pressure. However, it is expressed in a collective form and may have more weight because it comes from another commercial organisation rather than individual (and less 'expert') consumers.

A related form of pressure stems from insurance companies, which have particularly suffered in the USA under liabilities associated with the asbestos industry (Robins 1990, p.165). Insurance premiums and liability risks seem to be rising in importance and cost. For example, reinsurance underwriters had by 1990 paid $388 million in settlement of clean-up costs following the spill from the *Exxon Valdez* in 1989 (*The CityLine* (Greenpeace newsletter) No.13 August 1990, p.4). Insurance costs have particular resonance for 'green business' considerations, because 'insurers will be increasingly vigilant about companies' environmental exposure' (Robins 1990, pp.165–6) and may instigate penalties to ensure improvement. Hence, Touche Ross (1990, p.16) claim that the environment may prove 'a future debt burden', although business and insurance companies, banks and investors have yet to implement this in their everyday thinking.

In the USA, the important developments are related to the Superfund legislation (see earlier), under which the Environmental Protection Agency has identified 27 000 possible clean-up sites at an average cost of $25 million per site. This could have substantial impacts on liable companies. Robins (1990, p.56) cites the case of a Swedish Nobel Industries subsidiary which was forced into liquidation when faced with $60 million in clean-up costs for a single contaminated site—the cost was equivalent to double its annual sales. Superfund also deals with criminal and civil penalties: in the 1988–9 financial year, the EPA assessed around $37 million in civil penalties for environmental law violations, found 50 individuals guilty of environmental crimes and imposed $1.4 million in fines and prison sentences of eight years in total. Attempts to implement a Superfund-like system in Europe have not yet succeeded (see earlier).

Reporting and validation of company environmental programmes may help to lower premiums by making responsible companies appear to be better risks, linking investment and insurance as sources of pressure.

Pressure from angry investors may force the SEC [Securities and Exchange Commission in the USA] to require some disclosure of the potential liabilities of publicly traded companies. (Robins 1990, pp.56–7)

This may also introduce a longer-term perspective, where the duty of care placed on companies remains long past their direct involvement through retroactive legislation.

As well as the pressures from investors and insurers, businesses are influenced by other businesses more closely connected with their day-to-day running. Suppliers and retailers have close links which can serve as channels for pressure. In some cases, retailers are more responsive to market demand through their closer contacts with the public, and so may encourage or even require manufacturers to make changes on the basis of environmental damage. Companies claim to inform their suppliers of their corporate environmental policy and therefore request that these manufacturers fall into line or adopt a similar policy. For example, in 1992 IBM issued environmental self-assessment guidelines to all its UK suppliers and began to survey the environmental performance of its manufacturing subcontractors (Hill 1992, p.11) and British Telecom has assessed its suppliers' environmental standards since 1991 (*ENDS Report* 1995, 242, p.28). Further, Hoechst Celanese developed new technology specifically so that it might continue to supply Coca-Cola with plastic bottles under its new requirements (Biddle 1993, p.154).

In 1991, the large UK DIY retailing chain B&Q challenged its 450 suppliers to develop environmental policies and management. In a four-page advertisement (e.g. *Guardian* 1993, 16 August, pp.11–14), B&Q listed the bases of its environmental policy and consequent actions, including its Supplier Environmental Audit (SEA) launched in 1991:

> We explained to [our suppliers at a conference] that if they wished to continue trading with us they would have to meet our requirements on the environment . . . Since the conference, it has been part of our Terms of Trading that each supplier should have an environmental policy backed by an environmental audit . . . At the moment, 30% of our suppliers comply with this condition and we are working with the remainder to achieve rapid progress towards this requirement.

Ten suppliers failing to meet B&Q's requirements were delisted by November 1994 and the company has been keen to publicise its commitment not only to delisting but also to encouraging its suppliers to remedy their environmental faults with B&Q's support (B&Q 1995; *ENDS Report* 1994, 239, p.3). It pursued this also by joining the WWF 1995 Group where companies pledged that all their timber supplies would originate in 'well-managed' forests by 1995. Again, this illustrates the links between NGOs and business, here to put further pressure on business up the supply chain. This seems to have been particularly important to B&Q because the deficiencies of its suppliers were perceived to be jeopardising the retailer's reputation.

> Without an internationally agreed definition of sustainability applicable to both tropical and temperate forests backed up with independent verification, [our suppliers' claims of sustainability] all lacked credibility with customers. This lack of credibility was

highlighted in the statistic that whilst 90% of B&Q's suppliers would or could not tell the company which country their timber came from, over 50% stated that the sources were definitely 'sustainable'. (Knight 1996, p.1)

All this material suggests that pressure from retailers and wholesalers reaches manufacturers, who are also the main targets of investment and insurance pressures. This reinforces the notion that it is manufacturing sectors which are at the 'sharp end' of environmental pressure and therefore they are the leaders in environmental change to their operations and communications (see Chapters 3 and 4).

SUMMARY

This chapter has considered how different sets of pressures affect business and influence its environmental activities. Throughout, I have stressed not only that these pressures vary according to company profile but also that they are mutually influential. I have further suggested that business can influence these pressures in turn through its activities, publicity and lobbying, and this is examined in the remaining chapters. There is still little research completed in this area, however, certainly not enough to permit definitive statements of the levels of influence from the different sources or their consequences in the long term. Despite this, the business publications appear to address themselves simultaneously to small businesses, large corporations, retailers, manufacturers, the environmentally conscious and the environmentally sceptical organisations, in the belief that the same mechanisms of environmental assessment and publicity will apply to all. It seems clear that as business environmental activities expand, they will attract more research notice and be subjected to more critical analysis. Until then, I can only qualify the generalisations made above by noting the differential impact that the pressures have on individual companies and sectors.

The rest of this book now takes up the response of business to these four sets of related influences and considers the different forms this takes. Permeating all is the continued realisation that it is commonly difficult to separate the reactive from the proactive element of business environmental activity. Many businesses cast their environmental activities in a positive, proactive light when they are clearly reacting to one or more of the key sources of pressure discussed. The complications this introduces will continually resurface as we progress through the levels of business change, communication, self-regulation and lobbying to pursue a clearer idea of the range and implications of the business response to environmental issues.

NOTES

1. And to show that the same surveys obtain different results across time as well, further complicating the picture, results from the 1995 survey by Environmental Policy Consultants of 188 'mainstream'

businesses were: 84% said UK legislation was the main factor behind their purchases of environmental technology; 68% cited EC legislation; 59% cited corporate environmental policies; 47% cited cost savings and 26% cited new business opportunities (*ENDS Report* 1995, 251, p.4).

2. Similarly, the Swedish Greens failed to be re-elected in 1991 due to what has been called 'green fatigue' (Rüdig 1992, p.1).

3. Although the Protocol dealt specifically with CFCs and halons at first and later also with methyl chloroform and carbon tetrachloride, business attention focused on CFCs because of their public visibility. They were also far more widely used in industrial processes than were halons.

4. Additionally 45 to 60% of adults were identified as the 'green consumer base'—sympathetic to but not yet buying green products.

5. And 33% also selected products for environmentally friendly packaging on a regular basis.

6. Consumers can also be portrayed by business as hypocritical because they previously enjoyed the consumerist advantages of products which they now condemn as environmentally damaging and therefore criticise business for producing (I am indebted to Martin Purvis for this point).

7. Other NGOs, such as the National Trust and the Royal Society for the Protection of Birds, have taken a more managerial attitude to the environment and have tended to avoid public confrontation with business when lobbying.

8. There is the implication that Greenpeace is now involved in the production of 8 million fridges every year and not the conservation or reuse of existing ones. In its defence, it argues that the 'greenfreeze' model is 39–55% more energy efficient than standard models (Greenpeace, *The Greenfreeze Story*, n.d.).

3

Analysing corporate environmental change

Having discussed the issues and sources of pressure relating to business environmental change in previous chapters, I want now to discuss some forms of response by companies. First, in this chapter, internal corporate responses and then, in Chapter 4, responses centred on communication with external audiences.

As part of examining corporate change, I want to consider the level of business change more fully, particularly how 'green business' has been analysed as a spectrum of activity rather than as a single concept. Commentators have developed different theoretical and classificatory frameworks for the levels of business change in response to environmental (and other) issues. It is useful to examine these frameworks for what they tell us about current change and about the ideals held by commentators for the future. I shall first demonstrate the range of activities encompassed by environmental changes within companies, before examining some of the classificatory frameworks that have been applied to business environmental change in detail. I shall then look at how these help us to analyse the contribution of reaction and proaction to business response to environmental issues.

BUSINESS ENVIRONMENTAL CHANGE

Business environmental change involves a number of activities within the corporate structure which are differentially prioritised by companies, sectors and commentators. It is worth outlining these briefly before moving on to consider the classifications developed around them. We can divide them into:

- technological and operational modification;
- information-gathering exercises;
- environmental management; and
- communication changes (Chapter 4).

The first encompasses a huge range of technical changes to products and processes, many of which are implicated in the cost savings attributed to environmental change noted in Chapter 1. For example, the changes from CFCs to HFCs and other substitutes in refrigeration operations would fit into this category (see Chapter 2), as would the substitution of recycled material for virgin material in manufacturing companies or in stationery used by commercial companies and the change to non-hazardous cleaning agents in chemical manufacture (e.g. cited by 3M in Smart 1992). However, such technically related changes are often very specific to the particular manufacturing processes and business operations of particular companies. Additionally, they do not necessarily tell us much about the manner of response and the level of reaction, particularly where they are not externally publicised. For these reasons, I will concentrate on the other three categories which are more generalisable across various sectors.

Information-gathering exercises to gain environmental insights and a more thorough knowledge of areas of good and bad performance are a key area of change. They are particularly advocated by commentators as a first step towards improving environmental performance, on the basis that 'what you don't know about, you can't manage'. There are a variety of terms for this type of activity, 'environmental review' and 'environmental audit' being the most common. However, there is some ambiguity about the meanings of each (Hill 1992, p.3). Strictly, 'auditing' is the verification of reported information so that 'environmental auditing' should mean the verification of environmental information provided in environmental reports or other documentation. However, this is rarely carried out at present because the mechanisms for identifying qualified environmental 'auditors' in this sense are not yet implemented (but see Chapter 4 for developments). A second meaning of 'audit' is a full investigation of costs and benefits in an organisation, which is carried out by an external organisation; a third is where such an investigation is carried out by an internal team; and a fourth is solely reporting information to external audiences (Gray *et al.* 1987; see Chapter 4). Hence, meanings vary from strict verification by independent assessors to less rigorous internal reviews and publicity exercises. The most commonly adopted meaning of 'environmental audit' at present is the third—an internal investigation—and is often referred to, especially where it is first carried out in a company, as an 'environmental review' for internal consumption. This is explicit in the often quoted International Chamber of Commerce's definition of environmental audit as an internal 'manage-ment tool' undertaken voluntarily. This definition has been cited approvingly by John Elkington (1990, p.12), Andrew Blaza of the UK's CBI (1992) and Burke and Hill (1990). In this internal sense, the audit serves to direct attention to the potential benefits from environmental action, which may be environmental or economic but which in either case promote business profitability. In line with the arguments about business and environment compatibility in Chapter 1, com-mentators have therefore characterised environmental auditing as essential to companies' environmental and economic survival.

> Environmental auditing will not be a sufficient condition for business success in the 1990s, but increasingly it will be a necessary one. (Elkington in CBI/ICC 1990, p.23)

Table 3.1   The benefits of environmental auditing according to the ICC

- Facilitating information exchange within the company
- Increasing employee environmental awareness
- Identifying potential cost savings
- Aiding training
- Developing emergency response information
- Developing databases for internal information and decisions
- Giving internal credit for good environmental performance
- Convincing external authorities that environmental reviews are being taken
- Improving the insurance situation

Source: adapted from Elkington 1990, p.25

The ICC suggests several benefits to be gained from an environmental audit (Table 3.1) and it is interesting to note that most revolve around internal uses, except where legitimation in the view of external authorities is mentioned. This suggests that auditing may serve to pre-empt restrictive external regulation of environmental disclosure. Oddly, little mention is made of the public relations value of auditing where environmental improvement, or at least openness, can be shown.

It was manufacturing industry which pioneered environmental auditing in the USA (Elkington 1990, p.10). Early examples of audits are found in Allied-Signal (1976) and Ciba-Geigy in Switzerland (1981), whose Head of Audit and Staff Environment said at a conference in 1990 that audits are important for internal improvement and external credibility (CBI/ICC 1990, p.47). Examples of the process undertaken in the UK are found in the oil industry (British Petroleum and Shell), the chemicals sector (ICI) and the detergent manufacturing sector (Lever Brothers and Procter & Gamble; Elkington 1990). Again, this reinforces the argument that it is the high-profile manufacturing companies which are at the 'sharp end' of environmental attention and therefore are most innovative in response, aided by their greater resourcing.

Often an environmental audit (or more correctly an environmental review) is closely tied in with the development and assessment of internal 'environmental management systems' (EMS). These describe the location of responsibility for environmental concerns within management and the workforce, and detail the documentation necessary to monitor and improve environmental performance in the light of the company's environmental policy (see Chapter 4). An effective EMS should therefore include the establishment of responsibility from top levels of management throughout the company as well as channels of communication, inventory mechanisms, monitoring procedures and periodic reviews of progress (see Welford and Gouldson 1993). In practice, this involves the creation of explicit environmental job titles or the allocation of environmental management to specific individuals in addition to their main job responsibilities. In some cases, companies have made such managers responsible directly to the board, seeking to publicise the priority which management places on environmental issues. A major UK DIY superstore chain, B&Q, has instigated posts both for a high-profile 'environmental controller' at the centre and volunteer for 'in store environmental officers' at individual stores, the latter adding environmental concerns to their existing responsibilities.

Thus, the development of information-gathering exercises and environmental management are closely tied together and are both involved in new regulation, namely BS7750 and the European EMAS. These regulations and the communication of this information in the form of environmental reporting are detailed in Chapter 4. For now, I wish to consider how these forms of internal corporate change on environmental issues have been analysed in the literature and develop further the ideas of reaction and proaction in this connection. This leads us to consider changes to corporate 'culture', i.e. the thinking, value systems, assumptions and accepted rituals that make up the foundation for practical activities of particular companies. For example, the acceptance and incorporation of environmental audits into a regular cycle of management development imply a change to corporate culture which embraces environmental issues. It is less clear how important these issues might be within a culture that is dominated by non-environmental concerns. This is one of the aspects that commentators have sought to elucidate.

ANALYSING THE LEVEL OF BUSINESS CHANGE

There seem to be almost as many classifications of business environmental change as there are academic authors on the subject. I shall take some of the most typical in the 'green business' literature to pick out the trends across the range and to contextualise some of my earlier comments about current levels of business response to environmental issues.

First, Welford and Gouldson (1993, p.56) describe a hierarchical model which firms should follow in order to progress towards specific environmental and ethical performance objectives (Table 3.2). Stages 1 to 3 constitute economic reasons for business to address environmental issues, namely decreasing costs and increasing profits. Most business environmental activity at present fits into these stages, which imply a strictly practical and economically driven need for change. The remaining stages posit an increasingly expanding view of what 'environment' might mean for business management, including 'cradle-to-grave' assessment and going beyond legislative compliance (stage 6), dialogue with the community and publication of environmental information (stage 7), ethical considerations and 'participatory arrangements' with the community (stage 9). These stages therefore imply a much deeper level of change and, more specifically, a cultural change to managerial thinking and practices whereby environmental considerations infiltrate all aspects and operations. Very few companies approach even level 7 and most are still at levels 1 to 3. The model looks descriptive, but in fact this approach is rather more

Table 3.2 Welford and Gouldson's hierarchy of environmental strategies

Level	Operational objective or strategy
1	Minimising cost
2	Maximising short-term profitability
3	Maximising longer-term profitability
4	Innovation in products and strategies
5	Recognising a total cost concept (beyond purely economic assessment of environmental issues)
6	Integrating the environment with all activities; cradle-to-grave assessment and environmental audit
7	Widening the environmental base
8	Planning for uncertainty and risk, e.g. emergency planning, global impact assessment
9	Holistic management, e.g. change in company culture, participation by stakeholders

Source: adapted from Welford and Gouldson 1993, p.56

Table 3.3 Welford's spectrum of greening

Strategy	Ideology
Add-on pollution control Technological fix	Reactive
Environmental auditing Integrated environmental management systems	Proactive
Product stewardship Auditing for sustainability	Ethical
Culture change Regionalism and cooperation	Explorative
The transcendent firm Economic and societal change	Creative

Source: adapted from Welford 1994, p.20

prescriptive because it suggests what environmental strategies should be like for companies ahead of what they are actually like now; the authors in fact describe the higher levels as what companies should *aim* for (Welford and Gouldson 1993, p.57). The model is also predominantly positive in its language and recommendations, progressing from practical matters, such as efficient corporate 'housekeeping' to keep costs down (stage 1), to broader changes to corporate culture.

Similarly, Welford (1994, p.20) provides a 'spectrum of greening' in 10 stages[1] (Table 3.3), moving from 'superficial change' to 'fundamental change' at the furthest level. Again, the earliest levels are the current focus of business environmental activity—pollution control and technological modifications to operations and

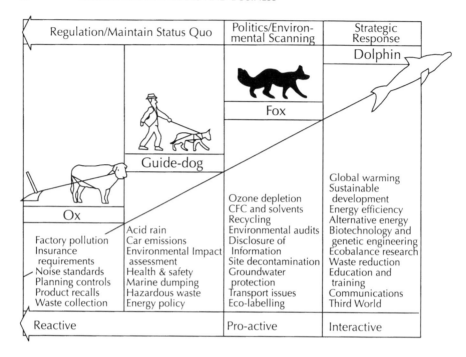

Regulation/Maintain Status Quo		Politics/Environ-mental Scanning	Strategic Response
	Guide-dog	Fox	Dolphin

Ox			Global warming
			Sustainable
			  development
			Energy efficiency
			Alternative energy
			Biotechnology and
			  genetic engineering
			Ecobalance research
			Waste reduction
			Education and
			  training
			Communications
			Third World

		Ozone depletion CFC and solvents Recycling	

Factory pollution
Insurance
  requirements
Noise standards
Planning controls
Product recalls
Waste collection
	Acid rain
	Car emissions
	Environmental Impact
	  assessment
	Health & safety
	Marine dumping
	Hazardous waste
	Energy policy
		Environmental audits
		Disclosure of
		  Information
		Site decontamination
		Groundwater
		  protection
		Transport issues
		Eco-labelling	

Reactive		Pro-active	Interactive

Figure 3.1   Elkington *et al.* learning curve of greening

Source: Elkington, Knight and Hailes 1991, p.230 (reproduced with permission)

products ('techno-fix'), but this time the categorisation seems based primarily on technological not economic factors. Similarly to the previous framework, cultural change is the basic requirement to move to 'higher' levels, being explicitly cited as a strategy (stage 7) and also being implicit in notions of stewardship, cooperation, transcendence and societal change. The positive phraseology of the 'ideology' column reflects the evangelical aim of much of Richard Welford's writings: whilst undertheorised, the thrust of this kind of material is clearly towards the positive characterisation of environmental change within a process of culture and value conversion.

This positive thrust is even stronger in the writings of John Elkington, a 'green business' guru of the 1980s and 1990s. In singly and jointly authored publications, Elkington has lauded the efforts of 'green capitalists', 'green business' and 'green consumerism' (see Chapter 2) in extremely positive ways, clearly exhorting environmental change rather than merely reporting evidence for it. A good example is given in Elkington *et al.* (1991), which is reproduced in Figure 3.1. Here, animals symbolise the character of companies of varying environmental attitudes. The graph progresses upwards as the positive message of environmental improvement is driven home: 'oxen' are goaded by legislators, 'dogs' are led by lawyers and 'foxes'

Table 3.4   The ROAST scale

Stage		Strategy/values
I	R	Resist environmental values
II	O	Observe and comply (with legislation)
III	A	Accommodate and adapt voluntarily
IV	S	Seize and pre-empt the environmental agenda
V	T	Transcend—holistic and proactive approach to sustainable development

Source: adapted from Welford 1994, pp.193–4

lobby politicians. The focus is on response rates and initiative as the key to environmental success, but the distinctions rest on cultural and managerial aspects of internal business operations, not economics, technologies or other changing external factors. This is also evident from the incorporation of an expanding list of environmental issues in the diagram, implying that as 'external' environmental issues multiply, a more proactive and 'interactive' response is needed. But the response that is ideal, the dolphin metaphor, is less clearly characterised than the others: dolphins are 'zeroing in on tomorrow's issues' and 'positioning themselves well ahead of their competitors', but there is little specification of how companies might achieve this, only the message that it will be fun because '[d]eeply intelligent and highly social dolphins learn fast and have fun in the process' (Elkington *et al.* 1991, p.231). Again, the current levels of business environmental response are linked hierarchically to the positive ideal which is as yet outside most business thinking and can therefore only be sketched, not fully described. Nevertheless, in a 1991 study, Elkington (with Dimmock 1991, p.18) tried to classify companies represented at a WICEM seminar by their response types, and claimed that 9% of representatives thought their company was 'reactive', 44% that it was 'proactive' and 26% that it was 'interactive'. This seems overly optimistic and is not necessarily applicable to business generally, in the light of the vague definitions of terms and the self-selecting sample of seminar attendees.

Even so, the frameworks discussed are typical of many in 'green business' literature. They are frequently prescriptive and deploy rhetorical rather than analytical language. This is nicely illustrated in Welford's use (but partial referencing) of John Dodge's ROAST scale (Table 3.4). This categorises firms' progress on the road to an ideal or 'utopian form of organisation' defined as the 'transcendent firm' (Welford 1994, p.192). As with the other frameworks, the two early stages apply to current business responses to environmental issues, being predominantly legislatively and economically driven (see Chapter 2). Stage A is more anticipative than R and O; stage S more proactive, implying an extended concept of stakeholders and dialogue with them; and stage T is well beyond that, requiring a firm that will 'proactively support and be responsive to all living things' in a 'near-perfect environmental performance reflecting the near-theological views of

Table 3.5 Layers of change in a company

	Layer	Possible environmental innovations
External	**Plans** e.g. information gathering	EMS, budgeting for pollution control
(amenable to change)		
↑	**Rituals** e.g. accounting	environmental auditing, environmental training
↑		
↑	**Assumptions** e.g. technological choices	valuing 'free goods' such as air, water; regarding resource inputs as finite
↑		
Internal	**Beliefs** e.g. ethics	morality of environmental protection
(resistant to change)		

Source: adapted from Smith 1993a, p.7

deep ecology' (p.194). Again, the later strategies imply cultural change as well as proactivity in the move towards a fully 'green' business, where management, ideology and even philosophy will have environmental foundations. Although the complexities of deep ecology are only superficially addressed,[2] the scale clearly implies exponential change between stages. Welford posits stages A and S as the current 'shallow ecology' of business environmentalism but stage T as full deep ecology. Yet, few businesses are at stage S—'seizing and pre-empting' the environmental agenda (p.194)—although they might like to be (see Chapters 5 and 6), and the further leap from S to T is more clearly an evangelical leap of faith rather than a theorised or practical undertaking (which may explain why no firms have yet gone so far).[3]

So the three frameworks considered so far emphasise economic, technological and external drivers for environmental business change at present. They suggest that in response 'green business' (and for Welford 1994, 'sustainable business') must address corporate culture more completely, especially corporate customs and thinking. This is a point made also by Denis Smith (1993a) where he considers the process of green change. At the shallowest level, environmental considerations are directed to planning and activities, such as pollution control and budgeting, and these prove the most amenable to change (Table 3.5). Once we move into aspects more central to company culture, change becomes more difficult. Hence, current 'green business' represents only a shallow level of change because these difficulties have not yet been resolved. The switch from modification driven by external factors to internal culture change is characterised by Smith as the gulf between changing activities, e.g. production processes, and changing rituals, e.g. accounting, and the assumptions underlying these, e.g. scientific and economic bases of production. The 'core beliefs'—responsibility and ethics—of the company are the most difficult to change, but would represent the deepest level of change if they were to be affected. Smith, Welford and Elkington, and their co-authors, therefore reflect similar opinions in that they see business change on environmental issues as having to address the transition from economic, technological and other externally oriented considerations to the more difficult level of fundamental and pervasive change to company cultures.

The problem with using these frameworks to understand the level of business change in response to environmental issues has three aspects:

1. As noted, the frameworks are often prescriptive, not analytic. Through their comments, they are seeking to influence business change, the subject of their commentary, in a mutually constitutive manner. Hence, company cultural change and the 'future' outlook of business on environmental issues are overemphasised at the expense of attention to the current situation.
2. The frameworks also move too readily from externally driven changes to culturally and internally driven ones. In practice, it is difficult to draw such clear distinctions because external and internal factors often operate in collusion (see Chapter 2).
3. They do not offer clear guidance on practical changes to fit each category, changes which could be monitored by external bodies to assess the level of implementation. This is a common problem with the 'green business' literature, which often divides culture and practice into separate realms and 'provides very little guidance on how to forge this link between the normative and the practical' (Corbett and Wassenhove 1993, p.127).

Because of these provisos, I want to take a simpler line in analysing the level of business change on environmental issues, relating it principally to external factors (detailed in Chapter 2) and the different styles of business response to these (detailed here and in Chapters 4, 5 and 6). First, I shall establish a less prescriptive basis for analysing the level of business change and explore the contradictions that it raises more fully.

COMPLIANCE, REACTION AND PROACTION

We need to develop a more robust and, preferably, simpler way of considering business environmental change, in order also to pursue the importance of reaction and proaction. Roome (1992) approached this when he classified business environmental strategies according to two external factors—the scientific significance of environmental impact and the public perception of environmental impact—yielding five environmental strategies for business (Table 3.6). Non-compliance, i.e. not responding to pressures from regulation, consumers and public opinion, is not a

Table 3.6 Roome's environmental strategies for business

Strategies	Action is
Non-compliance	passive due to costs or inertia
Compliance or 'legislation-push'	reactive due to legislation
Compliance plus or 'management-pull' driven from the top	reactive and proactive due to predicted legislation
Commercial and environmental excellence or 'management-pull' driven by all in the company	proactive due to value change in the business
Leading edge	innovative due to ethics

Source: adapted from Roome 1992

viable 'strategy' as such, because it will inevitably lead to business demise, either through decreasing profitability (non-compliance with the market) or prosecution or closure (non-compliance with regulation). Also, 'leading edge', i.e. practising 'the state of the art in environmental management' (p.19), is by its very definition an as yet infrequent strategy. Both serve to anchor Roome's scale at two extremes, but it is the central three categories which are most useful. These focus on the transition from reaction to proaction, driven initially by legislation and regulation to which companies respond and then begin to anticipate through research investments, public relations developments (Chapter 4), company change and lobbying (Chapters 5 and 6). Corporate values shift in response to take up some of the driving force by changing corporate culture and thinking to bring environmental considerations into a more central position.

We can compare Roome's framework with an even simpler classification, a threefold analysis of business change on environmental issues, drawing on Sethi (1981), Denis Smith (1993b, p.177) and N. Craig Smith (1990, p.59). Sethi's threefold classification (Table 3.7) is more useful than some of the more complicated frameworks because it avoids the prescriptive and evangelical leanings described above and, hence, does not overemphasise the 'proactive' category. It therefore allows us to place more emphasis in a classification on the current situation rather than the ideal position.

Sethi (1981) identified three levels of business responsibility which promote its social legitimation by paralleling social expectations. 'Proscriptive responsibility' is bound by external forces of economics and law, like the early stages of the frameworks considered above. Sethi's second category is essentially one of reaction and anticipation and incorporates the first category of compliance on which prescription must build. Here, the changes in society, e.g. public opinion, consumer

Table 3.7   Sethi's three categories of responsibility

Sethi's terms	Basis
A proscriptive responsibility to obey market forces and legislation	Compliance
A prescriptive responsibility to operate congruently with the norms and expectations of society and to be one step ahead of these	Reaction/anticipation
A proactive responsibility to promote positive change and to be responsive to society	Proaction

Source: adapted from Sethi 1981

demand and government regulatory strength, are monitored by companies and attempts are made to anticipate and pre-empt these where possible (see Chapters 5 and 6). Smith (1990, p.59) elaborates on these three categories to combine the arguments of profit and responsibility, and attributes business responsibility to the internal balance between corporate priorities of profit maximisation and responsible or proactive behaviour. Proactive responsibility is a more extraverted attitude where company activity looks to change society, e.g. public opinion, as well as its own operations and culture.

Sethi's three categories, developed for social responsibility, can be applied to environmental responsibility and linked to Roome's scale through abbreviating the three categories to 'compliance', 'reaction/anticipation' and 'proaction'. Compliance involves conforming with legislative requirements and economic pressures, i.e. doing enough environmentally to remain economically viable, and is clearly a minimum stance for business to survive in the immediate future. For example, IBM's 1990 environmental policy has compliance as its first objective (Cleaver in CBI/ICC 1990) and the ICC's audit definition (discussed in Chapter 2) functions on compliance. However, commentators have pushed for a stronger attitude to environmental change, based on one of two arguments. The 'leading edge' argument says that leading companies will advance beyond the minimal level of compliance because they see competitive advantage in so doing. The 'inefficiency of regulation' argument sees regulatory compliance as a slow, costly and uninspiring way to make companies change. Such commentators usually prefer either voluntary standards (i.e. proaction and self-regulation) or economic incentives (i.e. reaction/anticipation) rather than strict regulation, again emphasising movement beyond compliance as a key mechanism for change. A common and related argument revolves around the associated level of regulatory certainty: too low and businesses will not change because they risk anticipating changes incorrectly and therefore losing their investments, e.g. in the wrong kind of pollution–control technology; too high and businesses will never innovate because they will be too busy complying, leaving regulation as the only impetus and therefore requiring continual regulatory effort and upgrading (e.g. Kemp 1993, p.106). For business, an ideal situation would unequivocally identify those issues to be addressed but build in some uncertainty about means and absolute targets in order to maintain a degree of flexibility in implementation.

> Although excessive regulatory uncertainty may cause industry inaction, too much certainty will stimulate only compliance technology. (Ashford 1993, p.282)

These arguments are rooted in the preservation of business autonomy in the face of regulation. Hence, they see reaction as the best and most flexible strategy for business at present, involving the anticipation of consumer and legislative trends and the consequent organisational changes to meet them satisfactorily and enhance autonomy in the process. Du Pont's CEO (in Smart 1992, p.187) defined corporate environmentalism as 'an attitude and a performance commitment that place corporate environmental stewardship fully in line with public desires and expectations', thus emphasising the need to match and anticipate social change. This includes compliance but goes further by attempting to keep ahead of changes, whilst not forcing them through in a proactive manner. This embraces Sethi's (1981) 'prescriptive responsibility' and Roome's 'compliance-plus' strategy, and involves market research as well as developing policy and communication.

Proaction goes beyond either compliance or reaction/anticipation because it implies the active changing of the company itself, under its own drive and 'vision', not merely because of the unavoidable requirements of legislators or consumers. This is the most common use of the term in the business literature. However, it also implies the active influencing of external factors, including the market, the political context and the regulatory regime. The frameworks discussed so far focus on internal, 'corporate culture' change as a proactive, but intraverted, strategy. In a fuller sense, proaction includes the more outward-looking strategies involving changing society and legislation to help business match its internal operations (and possibly cultural change) to context. This sense has undoubtedly been downplayed in the business literature in order to downplay the power of business to influence society, something which has been analysed in other academic fields for decades (e.g. Galbraith 1972; Gorz 1989; Chapter 2). Some 'green business' literature is beginning to explore these issues (e.g. Welford 1994; Bebbington and Gray 1993) but it is clear that political environmental proaction by companies is not yet a key focus for analysis (but see Chapters 5 and 6 on political lobbying by associations).

Complications with classification

So how well does a threefold classification fit business responses to environmental issues? Chapter 1 noted that most business developments on environmental issues were shallow or 'reformist', i.e. bolt-on or cosmetic changes to normal operations, not fundamental changes to cultures or operations. Hence, 'companies are failing to convert their concerns into positive actions' (Touche Ross 1990, p.3) in the sense that most companies are still adopting an environmental strategy based on either compliance or reaction.

We should make a distinction, although not a hard-and-fast one, between small and medium-sized enterprises (SMEs) and large corporations. Commonly, SMEs are driven by compliance considerations and their greatest pressure is from regulators rather than customers, public opinion or other business sectors (*ENDS Report* 1995, 250, p.22). The larger companies have more resourcing to monitor, anticipate and react to changes more flexibly and rapidly. They also have higher profiles and are more responsive to customers and public legitimacy issues. The first three frameworks described above, geared predominantly to 'more than compliance', are unrealistic at present for SMEs, although they may be inspirational for larger companies. We can see this in the dominance of large companies in organisations and initiatives seeking to influence the environmental agenda (see Chapters 5 and 6) and in their pervasiveness as examples of 'good practice' (e.g. Hill 1992). The large companies also seek to establish their 'leadership' on environmental issues, in the sense noted by Roome (1992), as part of their competitive advantage and leadership of markets and other conventional business areas. For example, Du Pont's CEO declared that:

> the commitment [in 1989] to position us in the forefront of industrial environmental leadership has to come from within . . . The time was right to raise environmental affairs at Du Pont to a new level of conscious concern, to adopt a proactive means for dealing with environmental issues, and to evolve this new environmental stance into a source of competitive advantage for the company. (in Smart 1992, p.187)

However, most companies take a compliance or a reactive stance, so that we need a framework that explores these categories more than the proactive one which is the preference of the frameworks favoured by Welford and Elkington. But this becomes difficult because there are differences not only based on company size, but also because different environmental issues cause a proactive response for some companies and a reactive one for others. Also, company activities go on at several levels, requiring us to consider how far reaction, or proaction, is being developed at each one. For example, retailers might be seen as proactive in their environmental marketing, reporting and communication (Chapter 4), but less so in their governmental and political or legislative influence (Chapters 5 and 6). In contrast, individual manufacturing sectors might excel at lobbying and legislative influence in specialised areas of regulatory development. In each case, it is possible for (mainly large) companies to advance current practice proactively and therefore, according to the arguments in Chapter 1, gain competitive advantage. But there is little evidence of this being coherently done in the environmental field. For example, for market proaction it would obviously be simplistic to say that companies can force consumers to buy 'green' products and equally simplistic to say that companies have no influence on consumers (see Chapter 2). The reality is that companies must react and must also take proactive steps, especially regarding new products. Du Pont's CEO reflects this intertwining of two forms of response when he argues (in Smart 1992) that Du Pont was proactive in developing environmental management, but locates this in a response to changing public opinion, the Montreal Protocol and the publicised role of Du Pont as a major toxic producer via the US Environmental Protection Agency's Toxic Release Inventory. How then do we theorise the distinctions of compliance, reaction and proaction?

I would note that the environmental changes that these businesses make usually fit into a business-as-usual pattern, despite their efforts to assure their publics that changes represent an environmental overhaul of this pattern. The emphasis is thus firmly and explicitly on modifications rather than radical changes (see also Buck 1992). These changes are easy in that they involve little or no financial sacrifice, e.g. public recycling banks in retailers' car parks which are easy to administer and therefore require little financial expense to appear environmentally concerned. In operational terms, the substitution of products or part of products may also be seen to be easy due to the relatively low cost of research and development or the perceived onset of legislation requiring such change, and therefore cost, in the foreseeable future. Although voluntary as yet, changes may also be a premeditated response to incipient regulation and therefore represent anticipatory reaction rather than a proactive approach to the evolving environmental agenda.

Further, the level of change professed by large companies is often gradual, even rather ponderous, and made up of a number of steps or minor changes. This incremental approach is typical of a reactive but anticipatory strategy (Peattie and Ratnayaka 1992). It implies caution, due to the uncertainties of the demand from consumers for products and services and the uncertainties around regulatory developments (e.g. packaging, see Chapter 5). It also implies a longer-term view, a 'commitment' into the future, where minor changes steadily accumulate to represent major change. This incremental reaction to environmental issues is not surprising, even for 'leading-edge' companies, given the arguments in Chapter 1 that business may not be inherently compatible with environmentalism. Incrementalism has become the norm: 'because of the way environmental management has been interpreted by industry, it has increasingly become a defensive and piecemeal approach which does not break away from any of the dominant paradigms of traditional profit-centred business' (Welford 1994, p.82). Companies often discuss their own strategies as being 'proactive'. However, this masks essentially reactive attitudes because proaction is often described as being limited by external constraints, which differ according to the kind of proaction professed.

Business also seeks to influence the direction and strength of the wider environmental debate, through lobbying, research into legislative change and pre-empting legislation through demonstrating company cultural change, self-regulation and other strategies. Touche Ross (1990, p.3) cites four benefits of proactivity in that a company will be able to:

anticipate threats and develop plans to decrease their financial and logistical impact;
satisfy the likely future environmental demands of banks, insurers and investors;
identify opportunities to promote new markets and develop more business through
  environment friendly products and processes;
meet the 'sustainable development' demands of the Brundtland Report, and thus help
  to ensure the world's economic future.

It even sees company environmental policy itself as proactive (Touche Ross 1990, p.6 but see p.8), although we can also construe this mainly as anticipatory and defensive, given the arguments in Chapter 2. In these terms, proaction is a positive business move in the light of a rapidly changing political and regulatory climate. Hence, proaction is often exhorted by business commentators, such as Blaza (1992) and Carey (1992), so that firms can gain competitive advantages and so that business in general can avoid legislative compulsion and retain autonomy (Simmons and Wynne 1993). This leads to the downplaying of the effectiveness of regulation, as seen in Burke and Hill (1990, p.5), and the exhortation for industry to frame the debate.

> We hope to convince readers not only that the green agenda will enjoy top priority in the coming years but also that business has a key role to play in further defining and implementing that agenda. (Elkington *et al.* 1991, p.9)

However, the evidence for business proaction is confused, with some professions of proaction in public communications coupled with reactive and compliance-oriented changes in other areas, especially where regulatory developments are uncertain. Proaction will be dealt with in more detail in Chapters 5 and 6 where, specifically, its potential in influencing the wider environmental agenda is explored. For the remainder of this chapter, I want to suggest the complications of this distinction in practice, where ethics and company culture are linked to actions and external forces in ways which blur a clear-cut analytical classification. This will consider how far proaction, in the form of environmental responsibility, can be separated from reaction to regulation, economics and legitimacy problems.

BUSINESS RESPONSIBILITY AND THE ENVIRONMENT

The notion of environmental responsibility in business originated in earlier concerns about the social responsibility a company held for its effects on its employees and the surrounding community (e.g. Drotning 1972; Sethi 1981). The emphasis has now shifted to deal more intensively with the environmental performance, especially in the business literature, e.g. the Chemical Industries Association's 'Responsible Care' programme (Burke and Hill 1990; Simmons and Wynne 1993; see Chapter 5).

Gray *et al.* (1987) have related responsibility in an enterprise to accountability under the social contract. The acceptance of funds or other forms of reward by the agent (e.g. a company) from the principal (e.g. a shareholder, employee) requires the agent to discharge their accountability by providing information about the financial and other implications of those rewards. Financial accountability has long been enforced through required external audit procedures, but social and environmental responsibilities remain largely voluntary for business. There are as yet only informal arrangements for accountability on these issues (notwithstanding recent developments in BS 7750 and EMAS, see Chapter 4). Thus, although legislation operates in some areas of environmental impact, e.g. pollution legislation, the necessity for a firm to act responsibly and to report this publicly as part of legislative compliance is not enforced as yet.

The voluntary publicising of business responsibility has attracted much comment. Early commentators found it difficult to reconcile with the economic and profit-making priorities of business, denouncing it as 'fundamentally subversive' and inappropriate for business (Friedman 1988). Gray *et al.* (1987, p.10) term such critics 'pristine-capitalists' in that they wish the capitalist process to be kept pure and uncontaminated by intervention or moral judgements. In the main, however, responsibility and accountability are seen by business and commentators as necessary (rather than morally correct) because of the power and position of business in society (Epstein 1981), i.e. under the social contract mentioned above (Gray *et al.* 1987). Hence, environmental and ethical considerations should 'spring from the perceptions that industry's right to operate in a given community is not absolute' (Karrh, director of Du Pont, 1990, p.70). This transforms (big) businesses from purely economic operations into social institutions, which must bridge the 'legitimacy gap' (Sethi 1981) between their operations and the social expectations placed on them and thereby justify their existence and activities (Buck 1992; Smith 1990; Epstein 1981). Legitimation of business activities and responsibility is furthered by the implicit or explicit use of consumer sovereignty, in advertising and marketing, to justify actions which can be criticised on moral grounds (Smith 1990; and see Chapter 4).

In the wave of corporate environmentalism since 1988, business environmental responsibility has been widely used, both implicitly and explicitly, in marketing logos and publications. For example, the UK grocery superstore retailer ASDA labelled its own-brand products as 'environmentally responsible' in the early 1990s. Although some large retailers, e.g. Safeway, (Adams *et al.* 1990, p.137), have prioritised environmental issues when defining their responsibilities, more commonly the translation of business environmental responsibility into action may be very general or vague (see Chapter 4). The identification of environmental responsibility by a company is therefore a reaction to consumer demand and public opinion, but also anticipates (encouraged by market research estimates) that this 'green market' will continue to expand and become more concerned about business environmental activities. Anticipation is especially frequent amongst retailers, who are often regarded by manufacturers as closer to the consumer and therefore swifter in their response to consumer trends.

So, environmental responsibility is predominantly a reactive response to changing social expectations, despite its appropriation by business in a claim to proactive change in internal value systems. In either case, finding unequivocal evidence of the responsibility itself is problematic. Where the connection between the environmental benefits and the economic benefits of actions is tight, it becomes difficult to show whether the proactive environmental responsibility or the economic incentive stimulated the *avowal* of environmental responsibility. For some companies, the perception of environmental responsibility initiates an environmental review which must subsequently be rationalised and justified economically. Whereas the environmental benefit alone would prove insubstantial and not initiate change, alongside cost incentives it can illuminate unseen potential. This does not presume that the environment secures a higher priority than profit (Eden 1993b) because the two are mutually influential: in some cases environmental ideas illuminate a route towards cost savings, e.g. through increased efficiency, and in other cases cost incentives prompt actions which prove to have beneficial environmental consequences. In the latter case, it seems that in the UK the rising costs of landfill have driven forward reviews of in-store waste volume and handling by UK retail chains to reduce their environmental impacts, because 'by aggressively attacking recycling, there are opportunities to turn existing costs into revenue' (Marks and Spencer internal booklet, n.d.). Another major grocery retailer, Tesco, announced that its new recycling programme would save it approximately £12 million per year by recycling 160 000 tonnes of cardboard and replacing one-trip cardboard transit packaging with reusable plastic trays (*ENDS Report* 1995, 248, p.15). Again, this is a clearly reactive and anticipatory stance to changing external factors of economics and legislation, especially where the two bind together as in the proposed landfill levy in the UK which links packaging waste, recycling, tax regimes and integrated waste management strategies.

Where environmental responsibility has been promulgated, and especially where cost incentives form part of its rationale, it is necessary to consider whether this responsibility stems from internal or external sources. Is business culture changing proactively and internally as it claims, or is it a reactive response to external pressures? Companies tend to emphasise an internal source of environmental and social responsibility within the company culture. This is indicative of a declaration of 'culture change' by companies, typically generated by individuals in senior positions, e.g. directors and chief executive officers, who mould corporate ideologies (Goll and Zeitz 1991). This is expressed via 'vision statements' and other general (unquantified) environmental declarations (e.g Karrh 1990; and see Chapter 4). However, this internal identification of responsibility is contradicted where companies perceive their responsibility to be dependent on definition by consumers (via their expectations) and subsequent approbation (via their purchases). This twofold source was clearly and succinctly stated by a manager of Marks and Spencer (in interview with the author, 1992):

> We will do things which we think are morally and ethically correct, the sort of things which we think our customers will expect us to do.

Consumer sovereignty is therefore implicitly advocated as the ultimate legitimation. Public expectations define what businesses must do to gain public acceptability on ethical grounds (Sethi 1981; Simmons and Wynne 1993). The professed responsibility is an internalisation of social expectations and therefore a reaction to context not an internal development. This is made explicit, for example, in Cooper's foreword to Burke and Hill (1990) where, as the chairman of the Institute of Business Ethics, he gives four reasons for business to act responsibly: it is morally right; it can improve profits; it improves recruitment, respect and public relations; it offers 'a huge and potentially lucrative commercial opportunity'. This intimately links internal and external factors. Yet work has demonstrated that the savings and benefits in terms of money and enhanced image, not least the avoidance of negative publicity, are likely to be far more important than internal sources of environmental concern.

> Waste elimination, resource recycling and energy conservation may be motivated primarily by economic interests, but protection of the environment is a by-product of such actions. This illustrates the economic and corporate benefit that can be gained through environmental protection. (Touche Ross 1990, p.10)

Positing environmental protection as a by-product of profitable action is hardly the same as action being driven by environmental concerns. There are also many external pressures which force or encourage businesses to adopt environmentally responsible behaviour, but which are less tangible than profit or returns on investment (see Chapter 2).

The obvious corollary to the identification of this external source of responsibility is that when public expectations rise for individuals, they will also rise, although perhaps to a lesser height, for business. (Of course, if expectations weaken with the rising priority of economic costs, standards of acceptability may become less stringent.) Business will have to change to become morally acceptable, to live up to more exacting standards in order to gain public legitimation, and therefore will have to anticipate and react continually. Hence, business commentators exhort business to embrace proactivity, rather than lag behind social expectations, and to continue to redefine these expectations in its own terms.

It seems that business environmental responsibility is therefore essentially reactive, being prescribed by the context of public expectation because of the need for legitimation as well as by regulatory activities (see Chapter 5). This responsibility will include the obligation to obey economic and legislative principles (i.e. compliance). This fits with Hoffman's (1994) argument that corporate environmental practices are not directed at enhancing profitability but at establishing legitimacy (also Schjaer-Jacobsen and Bundgaard 1994). However, one of the problems for business of establishing legitimacy in environmental fields at present is that new mechanisms to validate their activities are not yet fully implemented, e.g. BS7750 (see Chapter 4). Such developments are necessary because the visible institutionalisation of responsibility operates as a source of legitimation.

A prominent example of responsibility and environmental change in response to the need for legitimation is shown in corporate response to the 1987 Montreal Protocol on CFCs and other ozone depletors (see Chapter 2). The CFC issue was really the first major non-food environmental issue to prompt a response from UK retailers (Simms 1992) and has now become so standard that it no longer exerts a strong stimulus for change. Although positive, it represents movement in the direction of incipient regulation and public opinion, necessitating change for all companies (Dudek *et al.* 1990), rather than proactive or internally driven change. However, the reality of compliance is downplayed in companies' portrayal of their anticipatory and environmentally responsible policies, e.g. from a 'quality' grocery superstore retailer in the UK: 'Sainsbury's feels a very real sense of responsibility to do all it can to stop using CFCs in its business' (Sainsbury's leaflet 1989; also 1991; see Chapter 2).

Although anticipatory change is pre-emptive, it downplays the external pressures on business to change by claiming only proaction. It offers a short-term competitive advantage until the other companies catch up with the legislation. It is therefore good for short-term profits and long-term business reputation, again demonstrating the closeness of tangible and intangible benefits of environmental change (Chapter 1). This is also seen in relation to retailers' perceptions of forthcoming EC legislation on eco-labelling (see Chapter 4), compulsory environmental audits and statements (see Chapter 4) and restrictions on packaging (see Chapter 5).

> Companies that can anticipate rather than simply react to these developments [in EC and international policy] will have a competitive advantage. (Burke and Hill 1990, p.5)

Companies themselves foster ideas of culture change as proactive and responsibility as internally derived. Even where companies are merely ahead of legislative requirements or market demand, they ascribe themselves 'pioneering' status: 'we've acquired a growing reputation for being proactive, rather than reactive' (Safeway leaflet 1991). But the proaction is limited: it is not up to companies to make moral choices *for* their customers. For example, Sainsbury's claims that it offers environmentally friendly products 'as an alternative to standard products to give customers a choice—"environment friendliness" can involve a price premium, or some limitations in the product's function, and the Company does not wish to impose these restrictions by not offering an option' (Sainsbury's environmental factsheet 1991, p.5).

If companies abrogate their moral responsibility to customers, they construct business as environmentally *amoral*. They pass full responsibility to consumers who, as individuals with a free(r) choice under consumer sovereignty (see Chapter 2), can define responsible actions and products *for* business (and see Rolston 1989, p.172). Similarly, Gorz points this out when discussing capitalism's incorporation of economic rationality:

> It was no longer a question of good or evil but only of correct calculation. 'Economic science', insofar as it guided decision making and behaviour, relieved people of responsibility for their acts. They became 'servants of capital' in which economic

rationality was embodied. *They no longer had to accept responsibility for their own decisions.* (Gorz 1989, p.122, emphasis in original)

So, the decision over what is morally acceptable is implicitly passed on to the consumer because of the need for legitimation (Harte *et al.* 1990; Smith 1990), and hence survival, in the public domain. The public become the ultimate repository of responsibility for moral legitimation of business actions (Eden 1993b). This contrasts sharply with companies' promulgation of internal responsibility and suggests that it may be more geared to image and symbolic action than policy statements indicate (Simmons and Wynne 1993). Responsibility is reactive whilst public communication uses proactive language to gain legitimation of business change. This demonstrates the contradictions between proactive responsibility, which is internal and often devoutly expressed, and reactive responsibility, which is external and contextual. Moreover, the level of responsibility and the consequent level of change depend on the issues. A single company might be compliant, reactive, anticipatory and proactive on different environmental issues, depending on their centrality to its own business, their public profile, related legislative interest and the ease of technological response.

SUMMARY

This chapter has shown how prescriptive classifications have encouraged business to be proactive in its response to environmental issues, but also that business is currently rather more predominantly reactive. Even the profession of proaction by companies and their embracing of environmental responsibility does not guarantee actual change. In a survey by Touche Ross (1990, p.5), half of the companies could cite no specific person who was responsible for environmental activities and impacts, despite the wide use of corporate environmental policies in the 1990s (see Chapter 4). Shallow changes are the norm, in the name of regulatory pre-emption and public legitimation as illustrated in the previous chapter.

So, having examined the pressures on business to respond to environmental issues and some of the ways in which we might classify these responses in corporate strategic terms, we can move on to consider the effects of these responses on various publics, including consumers, regulators and pressure groups. From intraverted ideas about corporate strategy, we can turn to the extraverted environmental activities undertaken by companies, associations and other business representatives. It is clear that, in the process of responding to environmental issues, business is also contributing to the development of those issues. It contributes to public opinion, political expediency and regulatory attitudes and to the framing of environmental issues in modern societies, and its influence should therefore be examined just as closely as these other elements in the environmental debate.

NOTES

1. The numerical labels used in the original do not correspond directly to the descriptive labels. I have chosen to use only the latter for clarity of discussion.
2   Readers wishing to go further should try Devall (1990), Dobson (1990), Eckersley (1992).
3. Although some authors make claims that they have, perhaps optimistically, e.g. Roome and Clarke (1994) on the 'transcendent firm'.

4

Business and environmental communication

Every company lives and dies not only in the market but also by its image. (Rolston 1989, p.166)

Previous chapters have looked at the development of and pressures for business environmental change and at the claims for reaction and proaction in corporate shifts. This chapter moves on to cover a second area of corporate environmental activity, namely public communication, which is more explicitly geared to the construction of environmental images and messages for external audiences. Such activities fall into two categories: *advertising and marketing*, a very public form of communication with a variety of audiences but principally customers of various kinds; and *company policies and reports*, which are not so widely publicised. Although geared to different audiences, these two forms both seek legitimation of company activities and involve reaction, anticipation and proaction, as introduced in the previous chapter.

Both types rest on the communication of 'promotional information', which includes advertising but also packaging, design and all levels of signification which represent, advocate and anticipate a company's products and services (Wernick 1991, pp.181–2). Hence, even company environmental policies, buried as they sometimes are in annual reports, are part of that process of signifying a particular company, of imparting a meaning to its customers, a meaning frequently hooked into environmental references, but certainly also into references to reliability, innovation, credibility, permanence, dynamism and other desirable qualities. This form of signification has become more important as self-service retailing has become more common and companies have been forced to reduce their reliance on personal service by shop assistants and therefore on the latter's ability to push particular products (Clarke and Bradford 1989). It is through promotion to wide, impersonally addressed audiences, often on a national scale, that companies seek to push products and corporate images. This is exemplified by national retailers whose brand names are not linked solely to specific products or product ranges, so that the latter cannot fully signify the corporate entity. As a proxy, many of these retailers produce consumer leaflets, for pick-up from shelf-mounted boxes, to convey product and corporate promotion.

Following on from corporate change in the previous chapter, I shall begin with the communication of activities through company environmental policies and reporting and the consequent building of a 'green' company identity. I shall then deal with advertising and marketing more generally, because these contribute to the company identity but often also dwell on products rather than processes and management and are also more publicly disseminated than the first form of environmental communication.

COMPANY ENVIRONMENTAL POLICY

The previous chapter noted how environmental responsibility, values and change have been emphasised by companies as part of a 'greening' of business. We can explore the same issues with relevance to communication here: how far environmental change is professed and publicly expressed through corporate activity. Business communication through advertising and marketing always attracts a great deal of attention, but the less publicised area of corporate environmental statements and reports offers plenty of material for analysing corporate communication. I shall begin with corporate statements as the general hook on which more detailed reports are hung.

A general 'green' image became increasingly important to companies in the 1980s (Higham 1990a), so that in a 1989 survey of business people 40% believed that a green image 'made commercial sense' for their company (Hilton 1989). The most common expression of this green image is the corporate environmental policy. Simply, this is 'a written statement of intent, implemented with a suitable system of management' (Datschefski 1992, p.5). However, the first part of the definition is more honoured than the second. It is therefore useful to consider how many companies have such a written policy and what the implications are if they do have one.

There is a great deal of discrepancy amongst surveys addressing the first question and estimates range from the very high to the very low. For example, a 1989 survey by Business International of more than 100 chief executive officers of leading European companies put the proportion at 48% and one by the CBI with PA Consulting Group put it at 75%.[1] More moderate estimates have been 40% (Coopers and Lybrand Deloitte in Carey 1992) and 29% (Touche Ross 1990). Touche Ross (1990, p.7) compared the environmental policy situation in the UK, where 29% of companies questioned had a written environmental policy, with Germany and the Netherlands where 100% and 50% respectively had a policy. It is interesting to compare this with Denmark, where very few companies had an environmental policy due to the already restrictive and comprehensive legislation there (Touche Ross 1990). In this case, compliance requires much more from companies than in the UK, so that companies were hard pressed to be proactive through appropriate policies.

In Burke and Hill's (1990) UK survey of 82 companies from *The Times* 1000 list, 37% had a written company policy and 54% had assessed their environmental impacts. However, higher-profile, more public issues had received more attention in these policies and assessments, revealing a selective attitude to the environment. For example, a greater proportion of companies had some company policy on the use of unleaded petrol (83%) and on CFC use (62%) than on recycled paper use and hardwood use (both 24%), catalytic converters (20%), reduced car use (15%) and biodegradable cleaners (12%). The more public profile issues were also those backed by government action or legislation—the preferential pricing of unleaded petrol and the Montreal Protocol on CFCs and other ozone-depleting substances. Furthermore, changes to using unleaded petrol and not using CFCs are primarily technical, involving substitutable chemicals or processes (Chapter 1), rather than the more fundamental changes that would be necessitated under a policy to reduce overall car usage, for example.

It is useful to compare these results to a survey of a different form by Harte and Owen (1991; also Owen 1992). They chose 30 companies specifically because they were seen as good disclosers of social and environmental information, in the eyes of the authors and of ethical trust managers. The proportion of this highly selective sample having a 'statement of corporate objectives' or ethical policy was 40%. This suggests that *far fewer* than this proportion across all companies, including those with poor ethical profiles and a lower degree of openness, will have such a policy. Many of the quoted estimates of environmental policy therefore seem high. Indeed, Datschefski (1992, p.5), of the Environment Council's Business and the Environment Programme, suggests from anecdotal evidence that less than 1% of companies have such a policy. This seems more probable than the other estimates if we are to include SMEs. However, much of the survey work has addressed only specific sectors or large firms (e.g. Burke and Hill 1990), leaving unproven the accuracy of such statistics across all businesses. A larger survey than those cited above, of 670 unspecified companies by Company Reporting, put the proportion having an environmental policy at only 5% (Macve and Carey 1992, pp.20–21).

We must therefore treat all these surveys with caution, because they are often selective in their sampling, either concentrating on large companies, which have the time and staff to look into environmental issues and the public profile to make this necessary, or receiving responses only from companies which already consider environmental issues and no responses from those who don't. In both cases the sample is unbalanced. A second problem is that the policy being counted in many surveys is merely the written statement and is not necessarily linked to internal changes or appropriate objectives, again representing only the first half of Datschefski's definition. In a 1993 survey by KPMG of 'most' of the UK *Financial Times* Top 100, 70 of the companies that responded had environmental policies but only 12 had quantified targets, 22 had carried out an internal audit and 14 an external one. Similarly in 1993, a survey by BMRB for the Department of the Environment in the UK found that 58% (of 507 companies with more than 200 employees) had a *formal* environmental policy but only 44% had a *published* environmental policy (*ENDS Report* 1994, 234, pp.12–13). This is a fine distinction: it makes one wonder how the policy that was not published might be disseminated through the company. There is clearly a great deal of difference between the public face of corporate environmental policy and its private operations (see *ENDS Report* 1993, 226, p.4).

As well as considering whether companies have a policy, we need to consider the quality of those policies. Overall, business environmental policies tend to be very general (Shimell 1991; Hill 1992, p.4), dependent on an overall 'commitment' to environmental considerations and often a 'vision' statement. For example, Karrh, a director of Du Pont, wrote that its policy does not specify 'what we will do point by point, but it describes how we will conduct ourselves' and generalises that 'performance excellence and ethical behaviour in environmental protection will be part of the company's core values' (Karrh 1990, pp.69–70). This is similar to the viewpoint of the Business Council for Sustainable Development (see Chapter 6) who note that:

> The issue is not whether the company vision looks good on paper, but whether behaviour and outputs change . . . it provides the framework and guidelines for stimulating action. Rather than a project to be completed so that we can move on to something else, continuously revising and focussing the vision becomes a major role of top management. (Schmidheiny 1992, p.85)

This has led even pro-business commentators to argue for more specific formulae (e.g. Burke and Hill 1990, p.17), to make company policies more comprehensive, rather than selective, and establish quantitative targets. However, although such quantification would make business environmental change more easily measured, it would also expose lack of change and increase public scrutiny (see fears raised about this in International Institute for Sustainable Development and Deloitte Touche 1992). Hence, quantified targets are rare at present.

In addition, Burke and Hill (1990) suggest that rigorous environmental policy requires other qualities. It should be communicated publicly (see next section), implemented and regularly reviewed, and developed democratically within the company through involving all staff. However, to ensure its comprehensive implementation, top-down management is advocated, using a central person as a repository for environmental responsibility. For example, in the DIY retail field in the UK, B&Q's corporate environmental policy is coordinated by an 'Environmental Controller' who organises all communications. Furthermore, policy should be geared towards a continual cycle of improving environmental performance (e.g. using environmental management systems, see Chapter 2) and therefore be dynamic (Datschefski 1992).

Table 4.1   The environmental policy of Nuclear Electric plc

General policy
The Company remains committed to reducing any adverse environmental effects of its operations to a practicable minimum.

Specific elements

Compliance	'Comply with all statutory requirements for environmental protection and, where appropriate, exceed these requirements.'
Minimisation	'Aim to reduce the environmental effect of its activities to a practicable minimum.'
Culture	'Increase environmental awareness and encourage commitment to good environmental performance throughout the organisation.'
Communication	'Develop communication with the public so as to foster a wider mutual understanding of the issues.'
Information	'Monitor progress and publish environmental reports with targets for improved environmental performance.'

Source: adapted from Nuclear Electric plc, *Environmental Report 1993–4* (1994, p.5)

Again, this ideal is rarely realised. It seems that policy statements are not indicative of deep changes in business, but constitute internal signals to staff and external signals to the public (Burke and Hill 1990, p.19), in order to legitimate companies and their environmental activities. Some examples of company policy illustrate this through their restricted commitment. In the UK manufacturing sector, the chemicals giant ICI has a policy which argues for a 'practicable minimum' of environmental effects, recognising that 'ICI activities, like all human activities, have some effect on the environment' but that the environment 'is, within limits, able to absorb certain man-made effects' (Burke and Hill 1990, p.27). In the retail sector, another leading UK company, Sainsbury's, makes a general positive pledge:

> To conduct our affairs with real consideration for the environment, both in the products we sell and in all operational practices throughout the business. (Sainsbury's press release, June 1991)

A more detailed environmental policy is given in Table 4.1, that of the UK's privatised nuclear power generating company, Nuclear Electric, as publicised in 1994. Like ICI, it focuses on the 'practicable minimum' of environmental impact and its policy further illustrates that, although compliance is a fundamental issue, given in the general and the specific policies, wider external legitimation is sought through changes to culture, communication and information. This is of course motivated by the poor environmental (and health and safety) image of nuclear power in the UK, leading in many cases to public hostility to the sector's operations and lack of disclosure. Hence, Nuclear Electric has pursued positive publicity through its extensive range of environmental policies and reports in order to counter criticism.

Generally, green corporate images also invite criticism about the greenness of the product ranges and other operations involved. For example, some of the biggest suppliers of fossil fuels in the form of petrol for domestic vehicles, which advertise unleaded petrol with strong environmental references and have coherent and well-publicised environmental programmes, may, simultaneously, have large stakes in world fertiliser production. Further, they may have been involved in the manufacture of pesticides, such as aldrin, which, although now banned, was manufactured in the UK into the late 1980s (Murrell in Irvine 1989b). Critics scorn this kind of partial reorientation:

> green capitalism can give exploitative, destructive industrial capitalism a veneer of ecological respectability. That makes it a potent propaganda tool for industrialists anxious to resist deeper change. (Porrite and Winner 1988, p.150)

A corporation's activities may thus be less green than its environmentally conscious and caring policies and advertising suggest, because it is attempting to attract public legitimation without making the substantial environmental changes that such promotion implies.

Environmental accountability and reporting

Despite the problems in ensuring accuracy in policies, their widespread adoption fed the increasing disclosure of environmental information, via annual reports and other publications, in the late 1980s (Harte and Owen 1991; Gray *et al.* 1987, p.93). Like its predecessor, corporate social reporting, environmental reporting serves to discharge accountability through external information provision and is hampered by recession and other financial constraints when financial accountability takes priority. In the dominant business paradigm, accountability is due mainly to financial shareholders in the company, but 'social accountability' extends this to include employees and communities around industrial sites, and 'environmental accountability' extends it even further to national and international publics (including NGOs) and the environment.

This extension is reflected in the current fixation with extending the concept of shareholders to encompass 'stakeholders' (e.g. Schmidheiny 1992, pp.87–9). For example, Miller and Quinn (1993, p.16) argue that the traditional concept of shareholders should be reformulated in terms of three groups of 'stakeholders':

- 'the environment' (which is rather poorly defined);
- shareholders in the financial sense and employees;
- other stakeholders such as local communities.

This would imply that corporate culture had to change to consider the impacts on these three groups during decision making and therefore this would affect the relationship between companies and all these publics. However, as with the frameworks in Chapter 3, this model is prescriptive rather than descriptive. There is as yet little evidence that 'the environment' is being prioritised above more traditional shareholder groups, given the usual rationale for 'green business' (see Chapter 1). More prosaically, Cannon (1994, p.137) suggests that there are now two groups of stakeholders: primary and secondary. The primary group includes financial shareholders, employees, customers, suppliers and government, i.e. those central to the running of business and those who regulate it. The secondary group includes the community, pressure groups, professional bodies, the media, academics and the law, i.e. the context or periphery surrounding business operations.

The extension of the core concept of shareholders in the 1990s to embrace 'stakeholders' implies a broadening of accountability and the adoption of acceptable ways to discharge it through environmental reporting. Reporting mechanisms to discharge financial accountability to traditional shareholders are standardised and regulated, but those to report on environmental accountability are only now emerging and gaining legitimation. As for company policies, for environmental reporting it is helpful to consider firstly its spread and secondly its quality.

Like company policy, reporting is often selective and the pattern varies. A few companies have produced full statements of their environmental performance across all sites and processes; others have completed environmental audits of varying intensities and published some results. For example, in a 1990 Coopers & Lybrand Deloitte survey, 29% of the small sample of companies had reported their environmental policies or performance in their annual report (Carey 1992), a channel used for public relations purposes as well as for financial reporting as required by law (Touche Ross 1990, p.8). In the survey noted above of 30 companies sampled for their good ethical performance, 40% had some general statement of environmental and social policy but demonstrated only limited environmental disclosure and this only in two parts of their annual reports, namely the (unaudited) Review of Activities and the Chairman's Statement (Harte and Owen 1991; Owen 1992). Gray *et al.* (1987, pp.60–61) compared US and UK reporting on environmental and social grounds. In the early 1980s, only 5 to 9% of the UK sample did any environmental reporting, whereas over 50% of the US sample already did in 1978. In Burke and Hill's (1990) survey of 82 large companies, 29% had set numerical targets for reducing their environmental impacts (especially on energy and waste), but only 15% had detailed their environmental performance in their annual report and only 16% had completed an (externally validated) environmental audit. Moreover, it is implicit in much of this literature that environmental reporting is dominated in its practices and principles by large companies. As we have seen before, SMEs have neither the resourcing nor the expertise to innovate in these kinds of ways, so that 'systematic environmental reporting is still very much a minority sport practised, on the whole, only by the larger companies in certain industries' (Gray *et al.* 1995, p.1).

The rationales provided for environmental reporting also vary. The main ones cited seem to be public legitimacy and competitive advantage, reflected in the trend that it is the largest companies which are furthest in the lead on reporting, due to their high public profiles and emphasis on competitive proaction. Of course, legitimacy and competition are interlinked because legitimacy in the public eye is related to legitimacy and reputation as a sound investment in the eyes of the financial sector. The emphasis placed on one or the other fluctuates according to circumstance. In a survey of about 70 companies in North America, Europe and Japan (Deloitte Touche Tohmatsu International et al. 1993), competitive advantage was cited as a reason for reporting by 18% of North American companies, 6% of European companies but none of the Japanese companies, whereas public relations was cited by 20%, 23% and 28% respectively. It seems therefore odd when the same study concludes that:

> Business leaders are clear about one thing. The main reason for reporting in the future will not be to position the reporting company as a responsible corporate citizen. It will be to secure a company's competitive position. (Deloitte Touche Tohmatsu International et al. 1993, p.8)

In other examples, environmental reporting is cited as important because it can gain credibility and acceptance and therefore ensure a continued 'licence to operate' in society (e.g. UNEP/IEO 1991, p.9; see Responsible Care programme in Chapter 5).

Like company environmental policies, company environmental reporting varies considerably in quality, displaying 'a bewildering variety of approaches' (Gray et al. 1995, p.1; also Fred Pearce 1992). It may be incorporated into a company's annual report as a separate section or as part of health and safety information. Some companies, again the large corporations and frequently those with connections to particularly polluting sectors, have developed 'stand-alone' environmental reports, which tend to be the most detailed form. Indeed, it is the large multinational companies which dominate the examples of environmental reporting in the literature (UNEP/IEO 1991, p.27; Deloitte Touche Tohmatsu International et al. 1993). An example of such detailed environmental reporting is the report on its 1993 operations by Novo Nordisk, the Danish commercial biotechnology company, its first stand-alone environmental report. This is interesting because the data were compiled by John Elkington's company Sustainability and therefore represent an external verification and validation of the Novo Nordisk report, albeit without reference to formal regulations. It includes quantitative data regarding outputs from 1990–3 for energy and water consumption, emissions to air of carbon dioxide, sulphur dioxide and nitrous oxides, and tonnages of various forms of waste discharged. It also includes specific targets, some qualitative and some quantitative, to improve on the present trends. In other words, the 1993 report sets baselines against which future improvements can be measured.

But this kind of stand-alone, independently verified report is not yet very common and there remains a great variety of environmental reporting across businesses. There are some similarities across the board, however. Companies often rely on 'specific narrative' (Owen 1992, p.68; Gray et al. 1987, p.93; Macve and

Carey 1992, p.39) in that they cite anecdotal evidence of their environmental activities. Although 'potentially auditable' (Owen 1992, p.68), at present this relates only to singular incidents not to company-wide operations, e.g. landscaping one particular site within a superstore retailing chain of over 300 sites. It is perhaps obvious that these specific narratives are as positive as they are selective, resulting in 'the general absence of bad news concerning environmental performance' (Harte and Owen 1991, p.9).

> 'Listening' is mentioned occasionally, but the value of saying 'I'm sorry' not at all. Whether such an attitude is the result of legal advice or corporate pride is unclear, but it is in any event a barrier to credible communication with critics. (Smart 1992, p.249)

As well as being positive, these narratives are described verbally rather than through quantifying environmental information. This contrasts with the quantification of financial information, which facilitates its comparison across companies and time periods.

> Some company environment reports still contain more pretty pictures of green fields and rainforests than hard facts expressed in graphs, numbers and league tables. (Fred Pearce 1992, p.21)

More quantified targets and specified means of achieving them have been argued for (Burke and Hill 1990, p.18; Elkington 1990, p.19), principally so that companies' serious commitment to such targets and the policies which prompt them can be exhibited. There are already a few examples of quantified and specific targets, but again they come mainly from (very) large companies. In the US manufacturing sector, 3M improved its 3P programme (see Chapter 1) in 1989 by specifically aiming to reduce its releases to air, water and land by 50% and its waste by 50% by 2000 (3M in Smart 1992, p.15). In 1989 Du Pont's new CEO set up a number of specific aims, including to reduce hazardous waste production by 35% from its 1990 level by 2000, to reduce toxic air emissions by 60% from their 1987 level by 1993, and to reduce carcinogenic air emissions by 90% from their 1987 level by 2000 (Woolard in Smart 1992, p.189). Nuclear Electric, in the example cited earlier, incorporates into its report a number of specific targets and marks those it has achieved and those it has not (Table 4.2).

Again, this is part of an attempt to publicise company openness through enhanced disclosure and emphasises how individual companies are voluntarily developing their own styles of environmental reporting. But this does not solve their PR problems because their reports are either unstandardised or unverified by some independent authority. Even with the publication of more specific voluntary targets, the problem of adequate monitoring remains: both the baseline figures and the yearly emission figures tend to be held by companies internally, except where their disclosure is forced by external regulatory agencies, e.g. for hazardous substances. Achievement of these goals cannot be validated by external bodies and therefore control remains with the company. For example, when *ENDS Report*

Table 4.2   Progress against voluntary targets by Nuclear Electric plc

---

Achieved three environmental targets, namely:
No site to exceed the ICRP doe limit to the public of 1 mSv/year
By mid-1993 purchases of equipment/plant to exclude CFCs and halons
Energy savings of 5% to be achieved for year 1993/4

On course to achieve nine environmental targets, namely:
All stations to implement BS 7750 and be ready for certification in 1994
Each station to identify environmental impacts and produce targets for improvements
in 1994
Each station to produce a local environmental policy in 1994
All non-operational sites to implement BS 7750 and be ready for certification in
1995
Company transport policy to be issued in 1994
Environmental training requirements to be reviewed in 1994
An environmental awareness package for employees to be introduced in 1994
A method for environmental performance assessment of suppliers to be introduced in
1994
Environmental assessment of main suppliers to be carried out in 1994

Failed to achieve three environmental targets, namely:
No site to exceed its authorised discharge limits
Water savings of 5% to be achieved for year 1993/4
Aerosols and solvents containing ozone-depleting substances to be eliminated by the
end of 1993

Source: adapted from Nuclear Electric plc, *Environmental Report 1993–4* (1994, pp.7–8)

investigated four major chemical companies' environmental reports in 1995, BP Chemicals admitted that 'almost 20% of the publicly reported reduction in total emissions from its sites between 1990 and 1994 was in fact due simply to better measurement' (*ENDS Report* 1995, 249, p.3). This had not been pointed out in the report alongside the publication of BP Chemicals' reduction figures. In addition, if the quantified targets relate mainly to disclosed information, there remains doubt about its validity: could the company have made the reductions solely through restructuring its organisation, such as selling off those parts of its business which were particularly polluting (see Fred Pearce 1992; *ENDS Report* 1995, 249, p.3)?

The problems are even more obvious for companies which have failed to achieve any quantitative targets which they set voluntarily. In 1990, ICI publicly set itself a number of itemised targets, especially relating to waste minimisation where it aimed to reduce releases to air, water and land by 50% by 1995. This initiative was in response to threats to ICI's 'licence to operate' in the face of growing criticism. Although ICI did manage to reduce its total hazardous wastes by 69% by 1995, it only managed to reduce its total non-hazardous wastes by 21%, therefore failing its overall target. In fact, its non-hazardous wastes disposed to land more than doubled in the same period, offsetting considerable reductions in those disposed to air and water. This was due to its neutralising of waste acid from the manufacture of titanium dioxide, which created over one million tonnes of gypsum per year. In other words, a by-product of dealing with its wastes in one environmental medium had become a 'new' waste problem in another (ICI *Environmental Performance 1990/95* 1996; *ENDS Report* 1995, 249, p.17).[2] It is this kind of predicament that has continually discouraged business from expanding its environmental disclosure in this intensive and proactive way.

Generally, environmental reporting tends to be reactive, a response to criticism over particular environmental issues, and has therefore developed mainly in manufacturing companies in a 'fire-fighting' manner (Gray *et al.* 1987, pp.87–93). It is also commonly shallow in depth and not all the environmental information is published. Companies prefer confidentiality to disclosure of information garnered (Elkington 1990, p.11; see Chapter 3), so that environmental reporting is rarely technically detailed or quantitatively presented (Macve and Carey 1992, p.46; also International Institute for Sustainable Development and Deloitte Touche 1992).

> Much of the disclosure appears to be linked to the development of an image, suggesting that it is good for both customers and shareholders that the company be environmentally aware, rather than representing a commitment to the concept of public accountability. (Harte and Owen 1991, pp.8–9)

Companies' adherence to motives of commercial secrecy obviously makes monitoring environmental reporting and assessing its quality more difficult.

> Support for environmental initiatives in the business community appears to be matched by a reluctance to release *detailed* information into the public domain. (Harte and Owen 1991, p.2, emphasis in original)

This is also revealed in the way in which companies discuss reporting. Their critics may discuss 'disclosure' but companies prefer the more positive and proactive implications of terms such as 'communication' and 'reporting' (Deloitte Touche Tohmatsu International *et al.* 1993, p.17).

Self-regulation of environmental reporting

It is clear that environmental reporting is being encouraged by business groups and commentators. Hence, given the above problems of selective, undetailed and unstandardised environmental reporting, attempts have been made to eradicate company inertia and establish reporting standards through voluntary schemes. These proliferated in the 1990s, but three key ones are worthy of discussion because they are frequently cited and moreover have developed similar criteria.

First, the World Industry Council for the Environment (WICE), an arm of the International Chamber of Commerce (ICC), addressed the problem. To fulfil one of the principles of the ICC's Business Charter for Sustainable Development (see Chapter 6), it published *Environmental Reporting: A Manager's Guide* (WICE 1994).

Table 4.3   Benefits of environmental reporting, according to WICE

Better **external business** including:
winning new business
increasing efficiency and therefore reducing costs and liabilities
encouraging investment

Better **internal management** including:
better information for managers
good management practice
better employee motivation
encouraging improvements

Greater **external legitimation** including:
retaining public confidence
responding to shareholders' and public concerns
providing 'evidence of environmental commitment'
gaining public credibility

Source: adapted from WICE 1994, p.5

This aimed to encourage environmental reporting, especially amongst smaller businesses, and to provide a practical guide to the content and rationale of environmental reports. WICE supports purely voluntary adoption of reporting and provides only an outline of possible contents in the hope of encouraging flexibility and the 'natural evolution' of practices (WICE 1994, p.3). They focus on four sets of contents:

1. qualitative statements of intent, similar to the 'vision statements' provided by CEOs on a range of issues besides the environment;
2. management systems and practices which review environmental performance;
3. quantitative information about environmental effects, including compliance information about how far regulations are met (or exceeded) and the consequent financial implications, e.g. for liability;
4. product information, including lifecycle assessments and recycling potentials.

Because WICE wishes the guide to be applicable in many different industrial sectors, it represents a 'broad-brush' approach rather than a detailed specification of reporting standards. Moreover, neither WICE nor the ICC monitors compliance with its guidelines.

Although the guide is, like most ICC publications, positive in its assessment of the potential for environmental reporting, it also notes the difficulties and supports full reporting rather than the type of selective disclosure which currently dominates practice. 'If your report is to be credible, publish good and bad news', it suggests (p.7). The guide also identifies three areas of benefit from environmental reports (Table 4.3), including greater external legitimation. Specifically, it considers the audiences for environmental reports, which will differ according to company type and activities, and emphasises the importance of gaining feedback and designing reports in order to deal with anticipated (and suffered) criticism.

Well-researched and well-presented information is a useful aid to dialogue and can help turn criticism [from NGOs] to understanding and cooperation. (WICE 1994, p.9)

Hence, whilst environmental reports have developed as a reaction to public criticism and awareness, WICE seems to be suggesting they have a strong anticipatory and even proactive function in establishing dialogue and pre-empting conflict.

Providing regulators with information gives them a wider appreciation of activities beyond legal requirements. The more they understand your enterprise, the easier it will be for you to influence the debate in your sector. (WICE 1994, p.10)

A separate attempt to develop guidelines for environmental reporting was undertaken by the Public Environmental Reporting Initiative (PERI). PERI is a group of 10 European and North American companies which wants, like the ICC, to encourage the voluntary adoption of environmental reporting. Like the ICC, it has developed in consultation with businesses a 'simple and non-prescriptive' set of guidelines for flexible use (PERI 1994). Its guidelines also aimed to meet the reporting requirements of anticipated and existing regulatory schemes, e.g. BS 7750 and EMAS (see next section) and the US Environmental Protection Agency's Toxic Release Inventory. They itemise 10 elements that environmental reports should describe:

- organisational profile;
- environmental policy;
- environmental management;
- environmental releases;
- resource conservation;
- environmental risk management procedures;
- environmental compliance;
- product stewardship;
- employee recognition;
- stakeholder involvement.

Like the WICE guide, the PERI guidelines are very general, suggesting that companies 'describe' and 'provide information' about the issues mentioned. They do suggest the provision of 'baseline data' so that progress can be monitored over several years of reporting, but mainly they provide issues on which to report, not specifications of the content of that reporting. They also stress that reporting can lead to 'constructive dialogue and cooperation' and better public awareness of company environmental performance (PERI 1994, Q & A sheet).

Thirdly, the Chartered Association of Certified Accountants (ACCA) published guidelines in the UK together with the CBI in 1994. These built on its environmental reporting awards scheme that had been running since 1992 and aimed to coordinate best practice into a coherent set of recommendations, especially for the members of the CBI's Environment Business Forum (see Chapter 5). In 1994, 36 companies submitted environmental reports for consideration and the main award winner, from Thorn-EMI, demonstrated the key aspects of reporting noted above, namely independent verification, comprehensive data coverage and a life-cycle analysis of each segment of its business (Gray *et al.* 1995, p.5). The guidelines were produced some time after the instigation of the awards scheme and replicate the four categories of information used by the WICE guide, listing similar benefits of voluntary environmental reporting including efficiency, public relations and enhancing investment. However, they stress the proactive potential when they note that environmental reporting within the CBI's Environmental Business Forum will help to 'gain a better political climate for the development of environmental legislation, leading to a sensible mix of regulation and voluntary action' (ACCA/CBI 1994, p.4).

This implies that reporting is a response to regulatory moves as well as to legitimacy needs, just like company environmental policy. However, unlike company policy, 'we find little or no financial market pressure (or encouragement) to discharge environmental accountability' through reporting (Bebbington and Gray 1993, p.6). Pressures for the development of mechanisms are therefore weak so that, in the continued absence of comprehensive regulation, the success of environmental reporting and its public legitimacy are debatable (Fred Pearce 1992). Its slow development is unlikely to be hastened except by tightened regulation and by the establishment of professional bodies to monitor and promote externally validated auditing of environmental reports. The Department of Trade and Industry in the UK did help to set up an *ad hoc* committee in 1994, bringing together a variety of groups to monitor green claims in non-advertising media, e.g. corporate reports, promotions and educational material (*ENDS Report* 1994, 231, p.27). Little has been heard of its activities so far except for some very general guidelines relating to 'integrity', accuracy and verifiability (DTI 1994).

External regulation of environmental auditing and reporting: BS 7750 and EMAS

As well as the three voluntary initiatives discussed, some regulatory moves have been made to deal with the slow and erratic development of environmental reporting. They include BS 7750 at the national scale and the EMAS regulation at the European scale, and I shall discuss these briefly to contrast with the self-regulatory schemes above.

The European Commission displayed an initial interest in environmental reporting and its regulation in 1989 and this prompted a variety of texts to demonstrate to companies how to set about an audit or internal environmental review (e.g. Elkington 1990, on behalf of the ICC). At this early stage, business was keen to pre-empt such regulation through establishing and encouraging environmental auditing across all sectors. I have already noted how, in many instances, auditing may serve merely as an internal review procedure, aiding compliance with environmental legislation and pointing out inefficiencies. The scramble for environmental audits was therefore reactive because it was seen as improving compliance, but also as anticipatory as the EC's legislation in this area was developing contemporaneously.

In 1991–2, the Commission developed its ideas for a Regulation for an eco-management and audit scheme (EMAS) which would involve environmental disclosure in company accounts, something specified in its Fifth Action Programme, 'Towards Sustainability' (1992). Piloting EMAS took place in 1992 (Hemming 1992) whilst businesses increasingly refined their own schemes so as to minimise changes when it was implemented and to pre-empt it if at all possible. The final Regulation (1836/93) was published on 29 June 1993 and entered into force in March 1995, the first companies being certified in the UK in August 1995. The scheme requires a company to pledge to establish environmental management systems and to commit to:

- drawing up company environmental policies;
- setting achievement targets;
- implementing environmental management systems to monitor achievements;
- auditing its achievements;
- reporting audits and having them independently verified (dependent on the establishment of a certification scheme for verifying bodies).

The overall emphasis is on a reiterative procedure of assessment to develop a continuous cycle of improvement through three-yearly audits. Rather than evaluating a company's absolute environmental impact, this means that audits will be 'a means for recognising efforts made to improve environmental performance, whatever the starting point' (Hemming 1992, p.9; and Welford and Gouldson 1993, pp.119–20). So, improvement is measured according to internal improvements regardless of environmental implications. The scheme has therefore come in for some criticism because 'the only requirement seems to be to demonstrate a capability for marginal environmental improvements within a self-determined framework of policies, targets, systems and assessment methodologies' (Welford 1994, p.75). Also, the entire scheme is based on *voluntary* participation by all businesses. The Commission originally intended to pursue mandatory participation but business lobbying successfully prevented this (Welford 1994, p.33). The Commission did, however, retain the right to adopt compulsory registration in future, adding power to the legislative impetus towards environmental audit. In the meantime, EMAS can be used as a positive marketing tool on letterheads and other corporate promotion, but not on products.

For business, EMAS will probably function as a legitimacy tool, providing independently certified evidence of environmental awareness and aiming to pre-empt public criticism and regulatory intervention. But it would also perform a similar job to environmental reviews and audits generally in improving environmental performance and therefore reducing operating costs, as noted above.

Experience suggests that both [EMAS and BS 7750] will help companies identify opportunities for greater efficiency and cash savings—while the longer strategic objective is to persuade environmental regulators that business is capable of getting, and keeping, its house in order voluntarily. (*ENDS Report* 1995, 247, p.18)

At the national scale, planning for the British Standard 7750 (Environmental Management Systems) began against the background of the first EC proposals for what became EMAS and drew on the previous standard dealing with total quality management (BS 5750). The main components of BS 7750 echo some of those already noted in the WICE, PERI and ACCA/CBI publications:

- a preparatory review;
- environmental policy;
- organisation and personnel structures;
- registers of regulations and impacts;
- objectives and targets;
- environmental management programme;
- operational controls and quality assurance on documentation;
- audits;
- reviews.

The standard was revised in February 1994 so as to be compatible with the final EMAS regulation and was then promoted as a credible, stand-alone form of self-regulation in the UK and as a way to step up to EMAS achievement.

Because EMAS and BS 7750 were simultaneously evolving, they became mutually influential. The EMAS environmental management system requirements draw on those of BS 7750 and the EC Regulation will be assumed to be met if national standards (e.g. BS 7750) are met where:

1. those national standards are recognised by the EC (as BS 7750 will be);
2. the certifying body is accredited in the member state; and
3. the environmental statement is published according to EMAS rules (*ENDS Report* 1993, 216, p.39; Gilbert 1994).

Business hoped that BS 7750 would match the general EMAS requirements, which it predated, thereby allowing companies a head start by changing to BS 7750 sooner and then using its format to argue for EMAS to match BS 7750 instead of introducing new (and possibly less favourable) criteria. The important difference between the two schemes is that BS 7750 does not have EMAS's commitment to the publication of audit findings regarding environmental performance, a disclosure with which companies are often uncomfortable (Welford 1994, p.74; International Institute for Sustainable Development and Deloitte Touche 1992). It has been suggested that BS 7750 would serve to introduce companies to the techniques, allowing them to cut their teeth on the less publicly scrutinised standards of BS 7750 before moving on to EMAS. The similarity between the two schemes should therefore encourage companies to set up an environmental management system and assess their progress 'before taking the key step to publication of performance. This allows a progressive approach, but with no wasted effort along the way' (Gilbert 1994, p.10). However, the impetus for environmental auditing has faded as recession has taken a firmer hold. The procedures towards implementation are progressing, but at a slower pace due to less pressing public attention since the early 1990s, such that companies have not been inclined to rush towards certification under BS 7750. Although it became available in March 1995 in the UK, by the end of that year only 39 companies had been certified, representing 'a depressing level of enthusiasm' (*ENDS Report* 1995, 249, p.2). Given the currently low level of participation coupled with their voluntary status, the impact of BS 7750 and EMAS is doubtful in the long run in terms of influencing business environmental activity.

> Moreover, we know from the experience of [BS 5750], that certificates of compliance are often pinned to the wall and subsequently forgotten about until the auditors are due again. There is no reason to suppose that the same will not happen with BS 7750 and the eco-management and audit scheme, and rather than provide for environmental improvement, that may lead to environmental negligence by the very firms which have been accredited by the 'highest' standards. (Welford 1994, p.79)

ADVERTISING AND MARKETING

Having considered internal corporate developments in auditing and reporting, we can turn to the more public communication of advertising and marketing. There is a considerable literature addressing these in general, in terms of their roles in capitalism and in society, their functions, forms and consequences. Much of this relates to, and often contradicts, the views put forward by companies and 'green' business literature as to how and why advertising functions. This is particularly important when we consider how advertising and marketing influence the green market and green consumers, thereby providing an important external pressure for business environmental change (Chapter 2). In this section, I shall explore these issues in general terms, before looking at how they operate in the environmental arena in the next section.

Advertising and marketing include not only advertising in broadcast and press media but also packaging information and design, product design, availability and distribution (Sinclair 1987). The advertising element has received the most attention, although many of the issues it raises equally implicate marketing as its sibling process. Advertising has been described as an instrumental process which has as its aim the driving of demand in order that products manufactured in modern industrial systems may be sold. Although primarily couched in economic terms, this basic construction can be broken down into two levels of reference—explicit and implicit—which demonstrate how promotion has permeated social and cultural realms (Wernick 1991; Clarke and Bradford 1989).

At the explicit level, advertising and marketing are about persuasion, in order that people buy certain products. Part of this process is product differentiation, whereby the 'unique selling proposition' of a product or service is promoted as the reason for purchasing it rather than its competitors. According to Sinclair (1987), this differentiation both establishes brand loyalty—benefiting the producer—and reduces the time needed for shopping—benefiting the consumer—because trusted products can be picked up without pondering the pros and cons of purchase. (Williamson (1978) explores this more thoroughly in her decoding of advertisements.) This frames promotion as information and adopts the language of consumer sovereignty (Chapter 2) because good information permits consumers to make the 'right' choices. In the environmental arena, imperfect information has been particularly criticised for encouraging people to make the 'wrong' choices, for example driving cars at high speeds (which is more polluting) and using throwaway products (tissues, packaged foods). But consumers frequently ignore or are passive in the face of (even good) information (Sinclair 1987), problematising the correspondence between consumer behaviour and advertising as information.

At a second level of reference, beyond the standard economic view of advertising and marketing, is a deeper analysis of its implicit function as deeply embedded in the relationship between consumption and production. In this view, advertising serves to create 'wants' rather than merely to inform people how best to fulfil their real needs (e.g. Galbraith 1972; Ewen 1976; Irvine 1989) and thereby obscures the appropriateness of consumption in favour of the diversification and expansion of consumption (Figure 4.1). This distinction between wants and needs gets at the basic issue of advertising: how far it creates demand or merely responds to it. If advertising is about creating demand, then it functions for the benefit of producers not consumers (Ewen 1976; Galbraith 1972), undermining claims to be purely impersonal persuasion. Ewen (1976) implicated the advertising industry as the particular agent of this promotion of consumer culture and both he and André Gorz (1989) criticise the creation of wants by business advertising to fulfil the needs of production rather than the deeper needs of people. So, this implicit function of advertising—to service production by guiding and pushing consumption—reverses the accepted marketing paradigm.

> Contrary to marketing ideology, markets do not already exist 'out there' in social reality, but are 'constructed'. (Sinclair 1987, p.97)

Both positions, consumer sovereignty and its rejection, are theoretically sound but practically too clear-cut. In reality, it is likely that both consumer sovereignty and demand management operate at different times and at different strengths according to sector. Unfortunately, we cannot easily distinguish these times and sectors in empirical evidence: it is only possible to establish that both arguments have some validity for different cases when discussing the issues raised by environmental advertising. (This reinforces the argument made in Chapter 3 that it is often difficult to untangle reaction and proaction in business environmental activity.)

Figure 4.1   Consuming more

Coupled to this implicit function, advertising serves to maintain the status quo, particularly capitalism, in the sense that it is 'socially conservative' (Wernick 1991, p.24; also Sinclair 1987, p.30). This means that advertising seeks to appeal to the conventional, to the majority, or in Wernick's (1991, p.43) pejorative phrase, the 'lowest common ideological denominator'. It portrays consensus and lauds conformity (Wernick 1991; Ewen 1976).

> Both in the values appealed to and in the symbols deployed, there is a deep bias towards the conventional and the most widely diffused. (Wernick 1991, p.42)

This leads to what Ewen (1976, pp.42–54) calls the 'massification' of the consuming public in order to facilitate production and marketing and reduce political and social change. This renders consumers passive, ready to be reformed under the 'culturally homogenising force' of commodification more generally (Wernick 1991, p.187).[3] Hence, advertising commonly advocates the purchase of more not less, which has implications for environmental damage due to increased production and use of resources and energy. The environmental implications of advertising are therefore potentially huge but largely untraceable in empirical terms. It further suggests that environmental advertising may need to make green references conventional or more orthodox before they can be widely employed, i.e. environmental usages need to be diluted in tone and implication or they will disturb rather than reinforce the cultural status quo.

Business has argued that environmental advertising is reactive to social trends identified by market research rather than an attempt to construct environmental markets. But, as noted above, the function of advertising is to persuade and in this sense advertisers and promoters are never neutral communicators (Clarke and Bradford 1989; Wernick 1991, p.42). However, Wernick (1991, p.25) suggests that the ideological function of advertising is not as conscious or intentional as the promotional (persuasive) one (although compare Ewen 1976), so we must look at the promotion to understand the ideology and not vice versa.

> Those who shape and transmit its symbolic material have no intrinsic interest in what, ideologically, that material might mean. Advertising is an entirely instrumental process. You promote to sell. (Wernick 1991, p.25)

This suggests that any social or value change because of advertising is an un-intentional outcome of that advertising, but this is debatable because, at ideological levels, the connections between advertising and economic structures become more difficult to untangle. Compare the views of Williams (1980, p.186): 'modern capitalism could not function without modern advertising'; and Sinclair (1987, p.33): 'advertising needs capitalism much more than capitalism needs advertising'.

It is, however, difficult to show (empirically rather than theoretically) that advertising achieves either its economic or its ideological functions:

> There is no necessary connection between the symbolic or ideological meaning of advertising and the behavioural responses which people make to it. (Sinclair 1987, p.63)

It is therefore possible that advertising's cultural effects are more significant than its economic ones in terms of directly influencing purchases, certainly over the long term (Sinclair 1987, p.31). In this sense, although advertising has some hegemonic potential, there is no guarantee that this will be fulfilled.

Environmental advertising and marketing

Histories of advertising in general (e.g. Williams 1980; Ewen 1976) illustrate how the focus has shifted, from initially promoting individual brands and products to promoting ranges of products and now to promoting the whole corporation (Clarke and Bradford 1989). As service sectors have expanded, promotion has grown to encompass not only physical products, but financial services, political campaigns and non-commercial organisations.

We can see this overall shift in miniature in the case of environmental advertising. By 1989, in a survey of City analysts, 80% of the respondents said that the marketing in their industrial sector had been affected by environmental issues (Hilton 1989). As marketing is traditionally more responsive to changing public consciousness than other company departments, environmental reorientation may begin in marketing and then feed back to other departments in the company (Elkington 1989). In the late 1980s, it seemed that marketing departments were 'in the front line of this shift in corporate thinking' (Hilton 1989, p.15)[4] and there was a proliferation of environmental promotions, including direct advertising, leafleting and product design and labelling. Initial interest focused on products with an environmental dimension, such as biodegradable detergents and mercury- and cadmium-free batteries, before moving on to product ranges, such as Reckitt and Colman's Down To Earth range of detergents, cars with catalytic converters and superstore chains' own labels, e.g. Sainsbury's GreenCare.

Advertisements for these products drew on a variety of symbols and meanings. The globe is a common motif, especially the 'blue and white planet in the black void' image used widely since the 1970s (see Cosgrove 1994). Other typical symbols include the colour green, wildlife references (e.g. to dolphins in the promotions for Reckitt and Colman's Down To Earth range) and tree or leaf references (e.g. in an advert for Siemens' mass transit systems). As well as global references, more local and 'English' landscapes and representations of the natural world are drawn on to anchor product and corporate identities in familiar and valued environments (Williamson 1978; also Daniels 1993). A further common trend is the use of children to symbolise future generations (in sustainable development, see Chapter 6): they appear in advertisements for Varta batteries, for Volvo cars and for energy-efficient household products and practices to reduce global warming, as produced by the Energy Efficiency Office in the UK.

As well as physical products, corporate images are now increasingly 'greened' through advertising campaigns. Investments, banks and environmental pressure groups have been promoted in recent years, again underlining the expansion of promotional activities into all areas of social and political life (Wernick 1991, pp.181–5). In the chemical sector in the USA, the Dow Chemical Company produced an advertisement in 1992 accompanied by a photograph of an ethnically heterogeneous staff supporting a toy globe—the copy discussed the 'environmental solutions that make a world of difference' that the company was developing. In the same sector in the UK, ICI's corporate campaign in the early 1990s focused on the theme 'World Class' and symbolised the control of famine through the (implicit) use of ICI products, presumably pesticides and fertilisers.

All these examples of environmental advertising raise arguments about the term's viability, similar to those raised in Chapter 1 about 'green business'. Within this range, environmental advertising falls into two groups:

1. well-intentioned advertising, where products are substantially different or modified and therefore promoted as 'green';
2. opportunistic advertising, where products have seen little or no change and are promoted as 'green'.

The former represents some degree of company change but the latter may not be complemented by any non-image change at all.

Some examples may clarify this point. Under the well-intentioned banner, we may include much recycled product marketing, especially where post-consumer and low-grade wastepaper is used, e.g. in Sainsbury's GreenCare toilet rolls. Under opportunistic, we may include the 'aggressive sales campaigns' (*The Ethical Consumer* 1989, 4, p.12) of many retailers, encouraged by marketers. The advertising and marketing magazines have attempted to aid this trend, and their treatments are illuminating. In 1989, *What's New in Marketing* suggested the following as food for thought for marketing executives:

> Packaging for *almost any product which can claim it is made up at least in part* of recycled materials based on a save materials, save waste pollution label. Paper products—from wallpapers to toilet rolls—produced on the same, recycled principle (*they do not have to be 100% recycled to give a green image*). (Fletcher 1989, p.25, emphasis added)

Much of the 'recyclable' labelling of products fits this category. Other forms of opportunistic marketing which attempt a green gloss without any other actions by the company are those in which no physical changes have been made to established products. In the early 1990s, labelling on bottles of Fairy Liquid dishwashing detergent proclaimed its contents to have been 'green' for over 50 years, and Tampax packaging claimed that its (bleached wood pulp) tampons have always been environmentally sound because they biodegrade after disposal. This sort of 'defensive advertising' (Irvine 1989a) is open to criticism as not only bandwagon jumping, but inertia and complacency as well.

> The biggest barrier to environmental change is denial that there is any problem with the environment or that there is anything wrong with 'business as usual'. (Datschefski, of The Environment Council's Business and the Environment Programme, 1992, p.10)

This clearly suggests that opportunistic advertising and marketing serve to legitimate currently unchanged business practices in the light of (new) public environmental concern. Products with little change are promoted by referring to the new consensus around 'green' values. This is assumed to be a conservative consensus, requiring only minor changes rather than major ones (Wernick 1991), but the change may be less than even the conservative advertising suggests.

> Most green advertising campaigns have therefore been cynical attempts to increase market share and profitability without any real regard to the principles of sustainability. (Welford 1994, p.16)

Individual products marketed under a green label often come from companies which produce other goods not labelled as environmentally friendly. This constitutes niche marketing: 'green' products are used as weapons in the commercial arsenal against competitors, rather than heralding a real change of emphasis for the company. Product differentiation is clearly at work, and is countered by groups such as those behind *The Ethical Consumer* and *New Consumer* magazines. These groups try to clarify the connections between companies and products so that consumers can make more informed comparisons, e.g. where environmentally marketed products are sold by companies with a poorer environmental profile generally or links to oppressive political regimes.

A related aspect is where advertising is exploited for its metonymic value, where a part of a product promotes a larger whole product or a series to which it belongs (see Wernick 1991, p.110). A clear example of this is unleaded petrol. One aspect of this product has been changed—its lead content—but this environmental connection is extended through promotion to colour the whole product, regardless of the environmental pollution caused by the burning of petrol and the production and running of its complement, the motor vehicle. Unleaded petrol pumps and lines are coloured green and there was a rash of green car stickers brandishing corporate logos when unleaded petrol was strongly promoted in the UK in 1988–90. This 'greening' of corporate identity stretches the environmental reference further to colour not only the whole product but the whole company.

Despite all the advertising interest, environmentally marketed products remain a minority. It is also difficult to measure their success because there are few types of goods bearing environmentally related labels for which there is information on their market share. Despite the large amount of market research on 'green consumers' (Chapter 2), statistics on 'green products' are hard to find. One example is the organic produce market, which in the UK rose in value from around £1 million in 1985 to around £100 million in 1990, representing 1 to 2% of the respective sectors (Hill 1986; British Organic Farmers *et al.* 1991; Lampkin 1990; Carey 1992).

Because of the difficulties in measuring product sales, and more so in measuring the sales by companies using environmental references for their corporate image, it is doubly difficult to point to the success or failure of green advertising amidst all the other influences which affect environmental concern. Robins (1990, p.116) compares two cases where companies used the WWF panda logo on their products—for Osram's energy-efficient light bulb, it bolstered and improved its market share; for Cipel-Mazda's 'green' battery, the company felt the move to be a failure. It is therefore difficult to say conclusively that promotion using environmental references is part of an *ideological* takeover of the environmental debate by business. For advertising companies (and their agencies), it does not matter whether people believe their advertising or whether it resonates with people's environmental values—what matters is whether it leads to purchasing, so that evaluation takes place on an instrumental basis (Wernick 1991, p.189). What is clear is that the business emphasis is on shallow or reformist environmental change. Advertising focuses on modifications of products and corporate images rather than political or social changes which might prove beneficial for the environment but which are difficult to package and sell for consumption. For example, it promotes as 'green' those batteries which are mercury- and cadmium-free but which still require 50 times as much energy to produce as they release during use.

Environmental advertising has the potential for proaction, for encouraging the public to approve and buy 'green' products. At present this is not realised because the emphasis is on slight environmental change to meet the shifting consensus of environmental values within the established pattern of production and consumption. The resulting promotion does influence this consensus, meaning that the public and the promotional versions of what constitutes 'green' are interwoven and continually renegotiated (Wernick 1991, pp.43–5). However, the process is not entirely driven by business. Neither consumer nor producer sovereignty operates unhindered: there is a mutual endorsement of consensus values between two sets of active agents (Wernick 1991, p.38). Yet again, this underlines the problems of distinguishing reaction and proaction in the environmental market.

Criticisms of green advertising

There are many consequences of advertising: the issues of consumer passivity and massification have already been discussed, and the relevant conflicts between business and environmentalist positions have been noted in Chapter 1. More important for environmental advertising are issues of technological fix, depoliticisation and commodification, issues which have generated a great deal of criticism.

Advertising seeks to create demands which can only be satisfied through purchase, not through unmediated or autonomous human relations. Hence, love and esteem are to be gained through using beauty products rather than through being a better person to those around us (Ewen 1976; Wernick 1991; Williams 1980). This necessitates a technological fix: needs are to be fulfilled through technical innovation and production, so much so that other ways of fulfilling needs are non-conformist and even subversive (Ewen 1976, p.213). All possible solutions are constructed in terms of commodities rather than lifestyle, social or political change. 'Advertising is the consequence of a social failure to find means of public information and decision over a wide range of everyday economic life' (Williams 1980, p.193), which, for him, represents the fundamental problem underlying all other criticisms of advertising.

Hence, advertising supports the dominant order again through the ideology of freedom of choice associated with consumer sovereignty, so that 'the social good is identified with private, individual consumption rather than public, collective production' (Sinclair 1987, p.27). Further, this argues that class issues and their associated inequalities are obscured through the overwhelming emphasis on consumption rather than production (Williamson 1978, p.47; also Ewen 1976). This tends to depoliticise the public because advertising, through arguing consumer sovereignty, redefines democracy as consumer participation in production and consumption (Ewen 1976). In other words, the possibilities of protest are presented in terms of consuming—not voting, demonstrating or lobbying. Economically constructed participation dominates (Hirschman 1970; Chapter 2).

All this implies that 'green' advertising will contribute to the commodification of the environment by emphasising a technological and depoliticised solution to our environmental ills. The application of environmental and 'natural' references to goods and services makes those goods and services our route of contact with the environment—we consume packaged 'nature' (Wilson 1992). In this way, economic values come to dominate cultural ones, so that the meaning of something is defined in relation primarily to its economic function. This would imply that environmental action is tied to economic channels through purchasing environmentally marketed products, investing in 'green' companies and donating to environmental NGOs. Political and collective action, such as petitions, demonstrations, voting on environmental issues, would be devalued (Hirschman 1970). Equally, the cultural and personal connection between people and nature would be undermined, so that people see the meaning of nature only when mediated through 'natural' products and never as itself (see Williamson 1978). Symbols and the economy which generates them become interwoven, so that 'what is promoted cannot be disentangled from what promotes it' (Wernick 1991, p.190). I would suggest that this process is far from complete on a cultural level but, for environmental issues such as deforestation, ozone depletion and global warming, the construction of action as purchase already forms the public consensus, supported by advertising, business and some aspects of campaigns by NGOs (Chapter 2). This points to the gradual commodification of environmental concern and to a further consequence of environmental advertising: that it takes the initiative away from environmental NGOs by allowing business to define environmental solutions (Chapter 1; Ewen 1976, p.220). Again, this suggests that business input to the environmental agenda is significant, but difficult to distinguish empirically.

One way in which NGOs sought to counter this was through exposing the often dubious early environmental claims of the late 1980s and denouncing 'green' advertising. In 1989, Friends of the Earth (UK) began an annual 'Green Con of the Year' award for advertising which misinformed the public about products or used inappropriate claims or descriptions (Verlander 1992). British Nuclear Fuels came 'first' in 1989, because of advertisements whch promoted nuclear energy as the cleaner alternative to fossil fuels due to its reduced emissions to the atmosphere (including, oddly enough, no CFCs). Its advertisements were strongly condemned by Friends of the Earth for 'factual inaccuracy, significant omission and for playing on the public's fears and ignorance about the Greenhouse Effect' (*Earth Matters*, 1989, 6, p.20). Other 'winners' are detailed in Table 4.4. Interestingly, Friends of the Earth stopped running the awards in 1993 because:

- the claims had become more complex and so were more difficult to dispute on clear scientific grounds so that the public could grasp the reasoning;
- the 'outrageous' advertising of the late 1980s had somewhat abated because companies had become more cautious;
- regulations for which the NGOs had been calling were beginning to emerge, e.g. on eco-labelling (see later).

Table 4.4   Green Con of the Year Awards made by Friends of the Earth in the UK

Year	'Winner'	Reasons given by Friends of the Earth
1989	British Nuclear Fuels	Use of the greenhouse effect
1990	Eastern Electric	Urged public to use more electricity to help the greenhouse effect
1991	Fisons	Argued that its peat-based compost did not endanger valuable British wetlands
1992	Meyer International	Claimed that its timber business had a marginal or beneficial environmental effect

Sources: *Earth Matters*, Friends of the Earth journal (1989, 6, pp.20–21; 1991, 10, p.7; 1992, 14, p.12; 1993, 18, p.6)

Together with the always meagre resources available to NGOs, Friends of the Earth decided that the awards had become a victim of their own success and that they could be discontinued because they had made enough of a dent in companies' consciousness and the development of other means to track claims became necessary.

Consumer distrust of green advertising

Criticism of environmental advertising has led to considerable distrust about it amongst the public. Environmentalist critics deplore the lack of information released by business about both its polluting processes and the effects of the products it promotes as 'green'. This issue resurfaces in the policy domain, where calls for public access to information and greater business accountability are frequent. A second problem for consumers is that, even where information is available, it is often technically complex or specialised. This makes it difficult for lay people to evaluate the claims being made by companies, which intensifies the consumer pressure on companies to clarify their positions and obtain legitimation.

Trust has to be earned by a business through legitimation (Zucker 1986), but there are as yet few regulated ways to legitimate business environmental informa-tion in advertising (but see later). Without these, consumers tend to distrust environmental information because they know that it is provided by business rather than because they know there are flaws in the information presented (Eden 1995). In other words:

> people perceive the message and the messenger as closely related. If the messenger is distrusted, the message may also be distrusted, no matter how accurate it may be. (Fessenden-Raden *et al.* 1987, p.100)

People perceive that business intentionally uses vague and misleading terms in order to sell products because, for the public, the undeniable motive of business to make money taints the information it gives and leads to doubt and suspicion of both motives and information in environmental advertising. Business promotional information is seen as manipulative, selecting and employing information to fulfil the needs of its providers.

So, distrust is produced from doubt about motives, not about facts, but this distrust then leads people into questioning the facts. The most obvious way in which information is seen as manipulative and therefore dubious is in its use of blanket terms such as 'environmentally friendly', 'ozone friendly' or even 'bio-degradable'. People see these as too vague to be meaningful and only motivated by the desire to make money out of green consumers. Such promotions tend not to be legitimate in the public view (Eden 1995). *Prima facie*, it appears that these labels are not useful because of the paucity of the information. They need to be more specific about the particular environmental attribute, to explain more clearly what the difference is between the 'green' and the 'non-green' option. In their vagueness, such labels only promote suspicion as they seem to gloss over the uncertainties perceived. This weakens their credibility and makes them appear intentionally misleading and a 'con', prompting scepticism that labels such as 'ozone friendly' are merely marketing hype.

People also feel that advertising claims are poorly regulated and checked, making it easy for companies to use such labels without further qualification. The company is therefore perceived as jumping on the bandwagon without fear of penalty and therefore nothing is taken on trust. This has made consumers very suspicious. In a 1990 poll, 49% of the public agreed that 'I do not believe labels that say products are environmentally friendly' (28% disagreed), and 67% agreed that 'saying a product is environmentally friendly is a way of getting you to pay more for that product' (17% disagreed, *The Guardian* 1990, September 14, p.33). So, more information might be a means of greater legitimation of 'environmentally friendly' labels, but in all cases trust resides not in the information itself, but in the provider of that information (Giddens 1990) and their motives.

Information may therefore be unintentionally or intentionally misleading, but people's evaluation of the motives of business leads them to deduce that the information given to them by business is intentionally misleading. In response to public distrust, therefore, companies attempt to legitimate their own decisions on the basis of their 'expert' knowledge and to neutralise public concerns through the presentation of technical assurances (e.g. Fischer 1993) to protect their environmental reputations. This has not, so far, countered public distrust of environmental advertising, which is hardly surprising given its early excesses.

Control of green advertising

In consequence, environmental advertising needs to be controlled to ensure public accountability and legitimacy. This can take two forms. First, advertising can be regulated by a legitimate external body under specific compulsory guidelines which therefore render advertising accountable and the process of examination transparent. Secondly, the industry can attempt to self-regulate, again under specific guidelines and transparent monitoring, but in this case through voluntary mechanisms with little external involvement. The second form is usually the first tried when advertising suffers a decline in public trust, e.g. in the USA in the 1910s (Ewen 1976, p.71) and in the UK between the 1930s and the 1950s (Williamson 1978, p.181). Voluntary self-regulation in the cause of legitimacy serves to pre-empt both public criticism and legislative moves. This allows the industry to appear active and yet address problems at the least injurious level, e.g. dealing with factual inaccuracy in advertising rather than exploitative or emotional uses.

We can see this kind of activity in environmental advertising, where problems of lack of information and lack of expertise led to increasing public mistrust of industrial 'greening' and scepticism of the motives of corporate environmentalism. The problem was partially a lack of standards. The late 1980s saw an upsurge in the use of terms such as 'environmentally friendly', 'ozone friendly', 'green', 'organic', 'phosphate-free', 'CFC-free', 'non-aerosol', 'unleaded' and so on, which was without legislative control and relied on self-regulation. This marketing free-for-all led to ambiguity and confusion, as environmental claims were made on shaky or spurious evidence or involved downright deception. Many of the marketing logos promoting the environmental friendliness of products resembled approved trademarks, and half the people shown such advertising ploys by the Consumers Association in the UK in 1990 believed that they were officially approved and met an independent set of standards (*Earth Matters* 1990, 8, pp.8–9). Such confusion had to be eliminated or at least counteracted as it both misled and disillusioned green consumers genuinely looking for less environmentally damaging goods and services. This was not helping business, which was selling the products, nor NGOs, who were orchestrating 'green consumer' campaigns, nor government, which was seeking environmental credentials. The public disillusionment and scepticism were widely recognised within industry and companies strove to dissociate themselves from it.

> The purely PR response is now widely taken for what it is—lip service. It does not wash with increasingly well-informed and sceptical stakeholders. (Datschefski, of The Environment Council's Business and the Environment Programme, 1992, p.4)

Hence in 1989, the Advertising Standards Authority (ASA), the industry's own UK body, took notice of complaints about environmental claims in advertising and produced a report addressing the issue.

> Some advertisers seem to be paying more attention to making sure their wares are perceived as sitting on the right side of the green fence than to checking the factual accuracy of their claims . . . advertisers have a special duty to ensure accuracy in an area where even the scientists are not absolutely sure of their facts. (ASA quoted in *Earth Matters* 1989, 6, p.20)

This 'special duty' prompted self-regulatory action by media associations such as the ASA on press advertising and the Independent Television Commission on broadcast advertising, because the advertisers needed 'to retain control of the regulation of their industry' (Buck 1992, p.37) rather than have regulation imposed as part of this backlash. Their codes of conduct were extended in 1990 to include environmental references, although environmental complaints made up only 1% of all complaints received (*ENDS Report* 1995, 240, p.30). The emphasis was clearly placed on the elimination of environmental inaccuracies and of terms such as 'green', 'clean' and 'environmentally friendly'.

- Claims should not be *absolute*, unless there is convincing evidence that a product will have no adverse effect upon the environment.
- The basis of any claim should, if possible, be clearly explained.
- The cloaking of claims in extravagant language should be avoided; this will only cause consumer confusion.
- Spurious claims should not be made.
- Advertisers *must* hold substantiation for *all* factual advertisement claims.
           (ASA/CAP guidance as given in ISBA 1992, p.3, emphasis in original)

The ASA decisions on complaints show that these 'relative' guidelines were held up only loosely, given the *ad hoc* nature of their procedures. For example, complaints were upheld about packaging from McDonald's not being fully recyclable; about Kyocera Electronics producing environmentally friendly printers; about the Vauxhall Corsa being the 'greenest car in its class'; and about Saab cars being able to clean city air. However, complaints were rejected about ICL computers being 95% recyclable; and about Lever Brothers' washing powder refill bags being 'kinder to the environment' (examples taken from *ENDS Report* 1990, 191, p.7; 1993, 216, p.24; 1993, 224, p.30; 1993, 225, p.27).

Although the codes urged companies to use relative terms and to be able to substantiate any factual (verbal) claims, they did not address the misleading portion of advertising, nor the visual and emotional appeals it often made using children and cuddly animals rather than environmental arguments. This type of advertising has been called 'unfair' (see Holder 1991) in its emotional not factual orientation. However, such an orientation has rarely been addressed by advertising watchdogs, in an implicit recognition that much product advertising necessarily plays on the fears of the public and thus uses emotional appeals as a key feature (Irvine 1989).

So, self-regulation met with considerable problems. In focusing on 'factual' material, the advertising watchdogs have found themselves out of their scientific depth in evaluating the basis of claims (*ENDS Report* 1990, 191, p.18). Also, there is little redress if a complaint against an advertisement is upheld—under peer pressure the advertisers may stop running it, but there are no financial penalties or requirements to publish apologies. In the UK, prosecutions have been brought directly by trading standards officers over environmental claims on packaging (not in mass or broadcast media) under the Trades Description Act 1968, but the penalties are still negligible. For example, Addis was prosecuted for labelling fire extinguishers containing halon-1211 'nonCFC-propellant, ozone friendly' and was fined £2800 plus £1300 costs (*ENDS Report* 1990, 191, p.18). The UK government promised in its 1990 white paper on the environment to legislate on environmental advertising claims, but has continually failed to do so. It later argued that this area was covered under the Trades Description Act, albeit not in a very clear way, and merely asked the National Consumer Council to consider whether any extra legislation was needed to clarify the issue (*ENDS Report* 1994, 238, p.22).

Moreover, in 1991, the ASA decided to stop monitoring green claims on its own initiative because it felt that they had become less misleading (*ENDS Report* 1991, 202, p.24). Between 1991 and 1994, the number of complaints about environmental advertising brought to the ASA by others halved from 192 to 89. However, this may not be necessarily because advertising was more rigorous in its environmental claims. For example, companies were complaining about each other when competitors felt that unfair environmental advantage was being sought. Lever Brothers complained about Reckitt and Colman's advertising for its Down to Earth range of domestic detergents which implied that the range was less harmful than those of its competitors (i.e. Lever Brothers) and the ASA upheld its complaint because it did not feel that Reckitt and Colman could substantiate the claim (*ENDS Report* 1994, 230, p.25). Companies had also begun to use the codes against their critics, e.g. the UK Timber Trade Federation and the Brazilian Embassy made a complaint to the ASA about UK advertising by Friends of the Earth which linked tropical timber operations with the deaths of indigenous peoples in its 'Mahogany is Murder' campaign. Their complaints were upheld by the ASA (*ENDS Report* 1995, 249, p.27).

In 1994, the British Codes of Advertising and Sales Promotion were reviewed to take account of those environmental issues that had emerged since their last review in 1988. They took a similar line to the ASA codes on absolute claims and extravagant language, but added two issues: that scientific uncertainty should be reflected in advertising, and that elements of products should not be said to have changed if they were never environmentally damaging, clearly reflecting the difficulties of claiming 'phosphate-free' on washing-up liquids which had never contained phosphates (*ENDS Report* 1995, 240, p.30).

Eco-labelling

Despite these voluntary schemes and adaptations, self-regulation did not eliminate consumer scepticism about disreputable environmental claims. Where self-regulation fails external regulation often develops, but this has not occurred in the UK for environmental advertising in the broadcast media. However, various forms of national legislation have addressed a related area of marketing-on-product promotion in the form of eco-labelling schemes, which should eventually influence the way in which broadcast claims are evaluated. Such schemes allow the use of a

Environmental fields affected	Product life cycle stages				
	Pre-production	Production	Distribution	Utilisation	Disposal
Waste					
Soil					
Water					
Air					
Noise					
Energy					
Natural resources					
Ecosystems					

Figure 4.2 Evaluation matrix for the European eco-label

licensed logo on products which have passed pre-set environmental performance criteria.

Individual countries (Germany, Canada, Japan) have had eco-label schemes for some years now and the German example, the 'Blue Angel' mark, is particularly well known, having been established since 1978 and now covering over 3000 products. This is often cited as the model for new schemes. The UK government accepted the need for a standardised eco-label in 1990, at the peak of 'green consumer' interest and during growing concern about environmental claims. The EC also set in motion the development of an eco-labelling scheme at that time. Its Regulation came into force on 13 April 1992, requiring member states to designate 'competent bodies' to look at eco-labelling criteria, product groups and individual applications for eco-label use and to conclude contracts if applications were successful. Although the criteria for inclusion should operate uniformly across the EU, individual member states are responsible for awarding the labels. The criteria are based on the results of a life-cycle analysis on the basis of the matrix shown in Figure 4.2, which quantifies the environmental impact of a product 'from cradle to grave' and identifies key areas where its performance contributes most to the overall impact. These stages can then be emphasised in the specification of eco-label criteria.

However, the process of establishing the criteria has taken far longer than originally estimated, due to disagreements between consulted groups as well as the slow operation of national bodies. The scheme was originally envisioned to be operating by the end of 1992, but the deadline was gradually extended to November 1993, when the first eco-labels were awarded to Hoover washing machines. Even after this, delays continued in the process of setting product group

Table 4.5   European eco-label product groups, February 1996

State of product group	Product group (member state in charge)
Addressed but criteria yet to be submitted to the EC	Packaging materials (It) Shoes (Neth) Cat litter (Neth) Ceramic tableware (Port/UK) Mattresses (Ger/Fr) Growing media (UK) Sanitary cleaning products (EC) Shampoos (Fr) Batteries (EC) Dishwasher detergents (EC) Converted paper products (EC) Floor cleaning products (EC) Rubbish bags (EC)
Criteria under consultation by the EC and member states	Insulation materials (Den) Fine paper (Den) Hairsprays (UK) Refrigerators (It) T-shirts/bed linen (Den)
Criteria agreed by member states	Light bulbs—double-ended (UK)
Criteria adopted and published	Light bulbs—single-ended (UK) Paints and varnishes (Fr) Laundry detergents (Ger) Washing machines and dishwashers (UK) Toilet rolls and kitchen towels (Den) Soil improvers (UK)

Source: adapted from *Ecolabel Criteria*, published by the UK Ecolabelling Board (9, February, 1996)

criteria, making progress very sluggish, with few product groups moving towards approved criteria (see Table 4.5). The scheme also seems dogged by conflict and lack of support. In autumn 1994, the EC proposed that the responsibility for the scheme should pass from it to the new European Environment Agency, because of conflicts between EC departments, especially DGIII (Industry) and DGXI (Environment). Member states reacted angrily to this and threatened legal action if the Commission did not press ahead. The Commission did so, setting swift deadlines for agreement on the criteria for eight more product groups by mid 1995, possibly hoping that these deadlines would not be met, thereby reinforce its argument to review the whole scheme (*ENDS Report* 1994, 240, p.29).

After over two years, only some Hoover domestic washing machines are so far sporting a European eco-label in the UK and, even for these products, environmental credentials took a 'back seat' in the advertising campaign (*ENDS Report* 1994, 229, p.24). By 1994, Hoover's share of 'up-market' washing machine sales rose 12% to 25% and the eco-label presumably made a contribution, but it is difficult to establish to what degree. Hoover's competitors were slow to follow its lead, waiting for evidence of tangible benefits before taking the plunge themselves. Hotpoint and Zanussi, although claiming to be able to meet the eco-label criteria, say that the costs outweigh the benefits (£500 fee for application plus 0.15% of subsequent sales) due to poor consumer awareness of the eco-label (a self-perpetuating argument). Few other product groups seem keen to emulate Hoover's lead so that, by the end of 1995, no other eco-label had been awarded in the UK and there was only one other company considering applying for an eco-label (*ENDS Report* 1995, 250, p.24).

There are therefore advocates and critics of the eco-labelling scheme. Advocates claim it will address the scepticism of consumers, thereby consolidating the 'green market', and offer an incentive for companies to develop new products to gain this marketing advantage. Environmentalist critics have highlighted the voluntary nature of the label: products will not be *required* to be tested and any that fail will not have to carry a negative label. This means that the label is a positive marketing tool rather than a regulatory strategy. Further, the influence of industry on the structure and criteria for the eco-labels has been criticised with the accusation that 'many ecolabelling boards have been captured by the very groups whose products are to be assessed' (West 1995, p.17). However, the level of business influence is disputed as DGIII (Industry) argued that industry's views were ignored in the development of product group criteria (*ENDS Report* 1994, 235, p.25). There seems little ability to compromise due to the ideological divergences noted in Chapter 1.

> Industry has generally complained that most of the proposed product criteria are too stringent, while environmental groups have claimed they are too lax. (*ENDS Report* 1994, 235, p.25)

Moreover, industry can threaten to withdraw from or even to veto the scheme, and therefore make it redundant, if the criteria and product group definitions do not meet its requirements. For example, the European Lighting Companies Federation have said that they will not be involved in eco-labels for light bulbs which address anything beyond energy efficiency criteria, e.g. packaging materials criteria (West 1995, p.19). The Scientific Committee on Phosphates in Europe, representing phosphate producers, published research stating that phosphates were not as significant in algal blooms as other parts of domestic sewage—it timed the report to influence discussions on the criteria for detergents, many of which currently contain phosphates (*ENDS Report* 1994, 238, p.10). Nationally, the British Aerosol Manufacturers Association has presented information to the UK Ecolabelling Board to demonstrate that non-aerosol (pump) sprays are not technically the best option for the hairspray eco-label (*ENDS Report* 1995, 242, p.27).

Business critics have also deplored the objective of European standardisation. Some companies feel that stricter regulations will confine them, whilst others already have stricter policies and feel that they will be penalised for being ahead of the standard. There are also arguments as to whether any comprehensive eco-labelling scheme would cause a stagnation in the development and adoption of new, less environmentally damaging processes. This assumes that the high costs of environmental protection would force smaller companies to continue operating older, less efficient and more polluting equipment rather than invest in new technology. Also, the more dynamic industries would be at an advantage as they would adopt pollution-abatement technology more rapidly during any expansions and modifications prompted by economic considerations.

The effects of these arguments depend on how far eco-label criteria are ahead of current practices: around 25% of products are likely to pass the criteria according to the Regulation (*ENDS Report* 1994, 232, p.28), although West (1995, p.18) suggests that the proportion may be much higher. Indeed, the British Paper and Board Industry Federation (now the Paper Federation) argued that the tissue criteria being developed in Denmark should be so defined as to enable 50% of current products to qualify. This problem is compounded by industrial concern that eco-labelling and other related regulations are 'potentially a highly effective form of protectionism' (*The Economist* 1990, 8 September, supplement, p.22), both within Europe and against non-EU states, and would allow those companies and countries which already produce less environmentally damaging products to guard their market share. Such issues are unlikely to be satisfactorily dealt with until the issues raised by closer European union are similarly resolved. In the meantime, the eco-labelling scheme has decelerated and has yet to have a major impact on the green market, especially as the UK Ecolabelling Board declined to use the £250 000 allocated to it to advertise the scheme to consumers until more than one manufacturer in a single product group had been awarded an eco-label (*ENDS Report* 1993, 226, p.25), meaning that advertising had only just begun in the spring of 1996 in the UK.

All this material suggests that once again the links between business and the environment are based on a combination of reaction and proaction. Lobbying activity on the eco-label has been reactive to and has anticipated draft legislation but is increasingly becoming proactive, with trade associations producing research in advance of their case. The argument for proactive advertising is a difficult one, partly because statistics on the 'green market' are so poor and thus give little empirical evidence for any argument. It seems that companies did certainly react to the increasing public interest in environmental issues, but did not react as quickly or as effectively when this soured and scepticism took over. This was where proactive self-regulation should have developed to deal with consumer changes, but it seems that the regulatory moves in Europe, rather than the consumers' mood directly, were the main focus of corporate activity.

SUMMARY

Environmental communication takes place through policies, audits and reporting as well as advertising and marketing. Such communication is essential to the image of individual companies and of business as a whole. This is because this image is in turn essential to gain legitimation from various publics—including NGOs, consumers and regulators—to pre-empt regulation, to influence it proactively and to maintain market viability. Much of this is often made explicit in business literature:

> Du Pont's environmental image is deliberately constructed to suggest that industry is capable of keeping its house in order without government interference. (Doyle 1992, p.90)

However, environmental communication as currently practised by businesses is poorly standardised and verified and is no guarantee of deeper environmental change. In some cases, it also serves partially to obscure this lack of guarantee in a cacophony of publications and publicity. It is again clear that regulation of communication is the main way to make it meaningful, but equally as yet there seems little political will to implement this against business opposition.

NOTES

1. This is disputable—the question was put to only the three-quarters of companies which rated the environment as very important or significant on a previous question. So the actual proportion responding that they had a corporate environmental policy was three-quarters of three-quarters, i.e. just over half of the total sample.
2. ICI justified this result of their waste management strategy by noting that, firstly, gypsum will increasingly be sold to other companies and hence no longer be counted as 'waste' and, secondly, that 'the environmental burden of the gypsum is considerably less than that of the waste acids which have been reduced by two thirds since changing our processes' (ICI *Environmental Performance 1990/95* 1996, p.5). This demonstrates its need for qualitative narrative even in presenting quantitative results.
3. There is a considerable amount of literature on the ideological content of advertising, relating to Marxist and structuralist analyses of its content, form and role, which it is only possible to mention briefly in this book (see Wernick 1991; Gallisot 1994; Williams 1980; Williamson 1978; and, for a theoretical introduction, Sinclair 1987).
4. But compare this with a paper by Steger (1993, p.155) who suggests, from his survey of German companies in 1991, that marketing was the primary function *least* affected by environmental matters. The consensus seems to be against him for UK business.

5

Business and the national environmental agenda

In previous chapters, I have looked at how business has taken environmental issues on board through reactive and proactive mechanisms in the realms of company culture, responsibility, advertising, policy and reporting. Although these were viewed as responses to external pressures, they were coupled with the anticipation of changing social and legislative trends as well as attempts to influence those trends proactively. In examining many of the topics, it was difficult to separate reaction from proaction with empirical rather than just theoretical accuracy.

To pursue this further, I want to turn to more explicitly proactive business environmental activity by considering the ways in which business seeks to influence the environmental agenda in its favour. Unlike in previous chapters, the activities under discussion focus on trade associations and other initiatives rather than solely individual companies. Large multinational corporations are undoubtedly important in influencing issues as well as in framing the reactions to them, through their lead as examples of 'greening'. Their influence is carried both directly to governmental contacts and indirectly through representation on associations and on committees in consultation with governments. However, much of their direct influence is privately conducted and not amenable to exposure in academic research, making it more useful to concentrate on the more public activities of the associations. Bearing this in mind, I shall first take the national case of lobbying for consideration in this chapter through UK case studies and then address the wider international scale of business activity in the next.

Business influence on the environmental agenda in the UK has two main incarnations: self-regulation and lobbying.[1] The second is an attempt to influence policy makers and other opinion formers directly and focuses on political and legislative issues. The first can be directed also at government, but in addition it seeks a wider role in establishing public legitimacy and thereby pre-empting government involvement in business activity. The two forms are mutually influential in that self-regulation often feeds lobbying by acting as a pre-emptive argument against new legislation, as we shall see. This chapter will therefore introduce the issues raised particularly by self-regulation and by lobbying on environmental issues, as well as the sources of business influence on the national environmental agenda in the UK.

SELF-REGULATION

Self-regulation is an important proactive environmental strategy for business in the 1990s, whilst often being equally prompted by reaction to environmental activity by other groups and seeking to earn government and public legitimacy. As seen in Chapters 3 and 4, public legitimacy can help to ensure company accountability and acceptability and, by relieving public pressures for business change, can affect government attitudes and legislative moves indirectly.

Self-regulatory schemes often begin with the development of a code of conduct through consultation within business. This code is then applied either to all business, e.g. environmental advertising and reporting (Chapter 4), or to the operations of a specific sector, e.g. Landfill Guidelines for waste management companies in the UK produced by the National Association of Waste Disposal Contractors (NAWDC). All companies to which the code applies are encouraged to adopt it formally and implement it in their operations or suffer appropriate penalties for its transgression, e.g. expulsion from an association. However, strict penalties are rare and the reliance is more on peer and public pressure than on legal measures. Hence, 'compliance' is rarely discussed in publications detailing self-regulatory schemes but 'commitment' is often mentioned. The difference is of course that self-regulation is voluntary and encourages change across a sector through collective movement, not disciplinary action.

There are two main rationales used for the development of self-regulation. First, such schemes are cited as pre-empting regulation by external bodies, the state and its agencies, e.g. NRA, HMIP, Trading Standards Officers. Importantly, self-regulation precedes direct regulation on an issue, but it is often prompted by the perceptible harbingers of such regulation in the form of draft legislation or sometimes public outcry over an incident. Self-regulation may be backed by the threat of government regulation or may be a complement to regulation—either way, it is often itself linked explicitly to existing regulations. For example, environmental advertising self-regulation (Chapter 4) both:

- follows the ethics laid down in established regulation but applies them in a new area; and
- experiments with and encourages new ethics and practices in advance of regulation (Boddewyn 1985, p.32).

'New' issues such as the environment are particularly prone to the development of self-regulation in the UK because 'the government has recognised the need for a new regulatory framework, but prefers to achieve this objective with as little direct government involvement as possible' (Grant 1987, pp.1–17), hence supporting and encouraging developments within business.

Secondly, self-regulatory schemes are cited as securing business legitimation from both the public and the government. This is often the case where there has been public criticism of a business sector or a publicised accident, such as the development of the Valdez Principles in 1989 after the oil spill from the *Exxon Valdez* off the Alaskan coast (e.g. Miller 1992). Here, the use of positive self-regulation in public relations can help to restore the lost public trust or at least attempt to contain it. Such a collective response is a proactive step to bridging the revealed 'legitimacy gap' (Sethi 1981) by demonstrating a rapid and coherent set of measures to remedy or prevent such incidents in future.

However, there are complications with self-regulation as a notion and in practice. First, even once self-regulation has been set up, government may still decide to legislate. So, government retains control by waving the 'stick' of legislation and 'the implicit threat is that regulation will be proposed and enacted if the industry does not police itself effectively' (Boddewyn 1985, pp.30–43). This may make self-regulation redundant in practice, although it is also possible that the enacted legislation will use existing self-regulation as a 'blueprint'. This would mean that the business viewpoint would at least have been incorporated into the external regulatory structure even if, in the process, the self-regulation were itself scrapped.

There are, secondly, significant problems of credibility. Public trust in the schemes must be built up for them to be accepted as alternatives to regulation, but this is notoriously difficult for businesses to ensure because they often sustain much lower levels of public trust than many other institutions (Eden 1995; Simmons and Wynne 1993). One way to attempt to build trust is to involve external individuals in the policing of self-regulation to enhance the credibility and legitimacy of the schemes.

> Outside members are very important . . . in establishing legitimacy. Their presence may not ensure that the organization is publicly accountable, but it does introduce an element of validation, giving the self-regulatory system a degree of public credibility and legitimacy. (Baggott 1989, p.451)

Thirdly, there is a more subtle problem noted by critics of self-regulation. Self-regulation has the potential to increase business power by becoming more widely influential than external regulation, such that business is able to define the political agenda and policy targets in the future. In this respect, self-regulation has been theorised as 'private interest government', which develops 'self-regulation through formally private organisations, empowered by a devolution of public interest' (Schmitter 1985, p.47; and Streeck and Schmitter 1985, p.13). Private interest government implements government policy by establishing it as the policy of business associations, which these associations then enforce in exchange for privileged access to government (Streeck and Schmitter 1985; Grant 1987).

State agencies . . . are often prepared to accept 'voluntary' collective self-regulation as an alternative to authoritative state regulation even if this implies certain substantive concessions and a loss of (direct) control on their part. What the state loses in this respect, it can hope to recover through lower implementation costs and higher implementation effectiveness. (Streeck and Schmitter 1985, p.20)

Self-regulation encapsulates the threat that power might devolve to those business associations administering the schemes whilst not ensuring that they become publicly accountable. This is part of Beck's (1992) notion of 'subpolitics', by which he characterises the recent policy input of science, the law and business as increasingly powerful but still peripheral—their supposed 'apolitical' status ensures their autonomy and non-accountability (and Grant 1984, p.12). In addition, Baggott suggests that government chooses self-regulation as a political strategy to maintain relations with particularly powerful groups, especially 'big business' interests, by devolving power to them in the hope of subsequent support in return. In other words, there are advantages for government in self-regulatory schemes as well as for business.

Self-regulation then tends to prevail where co-operation is essential for the implementation of government policies and where the regulated have a monopoly of technical expertise in a particularly important policy area. (Baggott 1989, p.444)

The effectiveness of self-regulation therefore depends on the public and political climate which defines its legitimacy and therefore its power. Self-regulatory and voluntary agreements between government and business dominate environmental policy in Germany, but are breaking down because of both 'the relative failure of private government in pollution control policy, and the growing distrust towards the political role of experts' (Aguilar 1993, p.238). In contrast, regulation in the UK has traditionally been discretionary (Vogel 1986; see Chapter 2), based on trust between regulator and regulated, so that 'Britain appears to be something of a haven for self-regulation' (Baggott 1989, p.438). Self-regulation is also compatible with the UK Conservative Party's ideology of free-market operations coupled with minimal intervention and deregulation (Baggott 1989, p.450). Therefore, the release of administrative and policing activity to 'licensed' associations in the UK, with the government retaining the threat of legislative action, may be seen as favourable particularly for environmental issues which remain peripheral to the overall ideological thrust of the current government. This devolution of power has increased business involvement in politics.

The paradoxical outcome has been the increased politicisation of business, aware of the advantages to be gained from cultivating relationships with government departments, or of the value of pre-emptive changes in business regulations. (Grant 1987, pp.179–80)

Where a self-regulatory scheme is endorsed by government, this grants prestige and privilege to the body administering the scheme, both in terms of public credibility and policy influence. But, in return, it also deflects public criticism away from government onto the self-regulatory bodies, without forcing the government to create and administer a lot of new regulations. This echoes Beck's (1992) discussion of the decentring of politics to a range of 'subpolitical' groups. However, this is only effective where those self-regulatory bodies are widely publicised and regarded as fully accountable, i.e. so that the public can monitor them, and this is only partially the case as yet.

From these criticisms, Baggott (1989, p.445) suggests that self-regulation is often simply a 'placebo policy' to 'organise' issues out of central politics, a manoeuvre which could insulate business from further government action. He argues that where industrial bodies are particularly powerful, they can then reject external regulation in favour of their own self-regulation (p.450). But in most cases, the 'stick' of external regulation and the tenuous business grip on public credibility serve to constrain the influence of self-regulation, so that:

> self-regulatory systems are institutions of compromise . . . This does not necessarily prevent them from being useful instruments in regulating powerful private interests and areas of great technical complexity. Indeed, they often appear to be the most appropriate instrument in such cases. But the most important thing is that they should be capable of being reformed. For it is when self-regulatory systems become insulated from external pressures that they become dominated by private interests and are rendered useless as instruments of regulation in the public interest. (Baggott 1989, p.452)

To explore the above considerations further, I shall consider two examples of self-regulation that have been stimulated by environmental concerns, amongst other things. Both are sectoral in scope but offer contrasts in coverage, strength and public profile.

The Packaging Standards Council (PSC)

The first example is of the Packaging Standards Council (PSC). This was set up in the UK in June 1992 by the Industry Council for Packaging and the Environment (INCPEN) as an independent body encompassing people from industry, science, trading standards offices and consumer organisations, and claims it has received 'enthusiastic support' from the DTI and the DOE (PSC 1993, p.6). It has three main aims:

- to be a consumer 'watchdog' to which consumers can complain directly about excessive packaging;
- to 'educate' the public about packaging, i.e. to encourage a favourable view;
- 'to encourage and promote good practice and to seek continuous improvement in all aspects—functional, social and environmental—of the packaging of goods offered to consumers' (PSC 1993, p.5).

To progress towards these aims, it has developed a *Code of Practice for the Packaging of Consumer Goods*, an essential first step in the generation of self-regulation.

Its origins in INCPEN, a UK industry association with membership comprising manufacturers, packer/fillers and retailers, meant that the Council risked being seen as merely a representative of sectoral interests and not an objective judge of the merits and demerits of packaging and its environmental impacts. Despite its receiving £100 000 from INCPEN in 1992–3, the Council decided to look for alternative sources of funding because it did not wish to be dependent on this single source (PSC 1993, p.16). Clearly the Council wished to enhance its reputation by distancing itself from the financial support offered by INCPEN as an industry body. Similarly, the inclusion on the Council of non-industry representatives was seen as valuable to its status in public and governmental eyes as an impartial judge of packaging. Establishing legitimacy in such a way is important for all self-regulation, as noted above. But, simultaneously, within industry the Council had to maintain a reputation as a body which, whilst providing an acceptable public face, did not constrain the performance of the packaging sectors, demanding a problematic two-dimensional credibility: a watchdog for consumers but a reliable friend to industry.

> The manufacturers and users of packaging are increasingly coming to see the Council not as an irritating, ankle-nipping self-appointed 'watchdog' but as a knowledgeable and professional clearing house for the exchange of valuable information between industry and consumers. (PSC 1993, p.2)

Despite external involvement, implicitly the Council is still not publicly accountable. Its composition is determined within industry and it judges complaints brought to it in an *ad hoc* way, according to its own criteria in its code of practice. The Council does not undertake full-scale monitoring of packaging performance across industry, nor could it afford to in any comprehensive way on its current budget. Further, it does not have any effective sanctions with which to penalise those trangressing its code, should they be identified. As with all voluntary self-regulation, a company can choose whether or not to adopt the packaging criteria and, if told by the Council that it is contravening those criteria, the company can choose to modify its packaging or to ignore the Council's recommendations as it sees fit. So the Council cannot penalise companies even if complaints about their products are upheld. The effectiveness of the Council in consumer protection and in forcing business change is therefore doubtful. Whilst companies can potentially gain some credibility by claiming to conform with the PSC code, the consumer is commonly unaware of the lack of rigour with which that code is applied and cannot therefore distinguish it from legally enforceable regulations, thereby causing confusion and disillusionment with claims (see Chapter 4 on advertising).

So, the Packaging Standards Council has some problems as an example of self-regulation. It is more a pre-emptive response to the developing EC Directive on packaging and packaging waste during 1992–3 (see later) than a strong piece of industrial self-policing. Further, it does not have a strong public presence and is unlikely to insulate industry from further legislation through representing a publicly legitimate and government 'licensed' alternative to the external control of packaging. Considering the subsequent developments in packaging waste management (see later), the UK government certainly does not seem to have regarded the Council as sufficient in this respect.

The chemical industry's Responsible Care programme

A second example of environmental self-regulation is the chemical industry's Responsible Care programme. Chemical associations are major players amongst business associations on environmental issues all over the world, reflecting the industry's position at the 'sharp end' of environmental criticism. In the 1980s, the international chemical sector's poor reputation, credibility crisis and consequent regulatory attention led the industry to consider initiatives to engender a more positive image (Dow Chemical Company in Smart 1992, p.71). The Responsible Care programme is one of the most well known, first developed by the Canadian industry association in 1984 and since adapted for different countries, e.g. for the USA by the Chemical Manufacturers Association (CMA). It was launched in the UK by the Chemical Industries Association (CIA) in 1989. The CIA represents over 200 chemical companies and is probably the largest sectoral association in the country, having over 70 staff and a strong lobbying profile. Adherence to its Responsible Care programme was made a condition of CIA membership in July 1992, but Responsible Care, like all self-regulation, is still in essence voluntary, as chemical companies are under no obligation to belong to the CIA in the first place. The programme aims at 'continuous improvement in all aspects of health, safety and environmental protection' (CIA 1992a, p.3) and the CIA argues that it is a 'culture' pervading a company, not just management-driven, top-down operational change (CIA 1992b). Responsible Care's guiding principles, to which CEOs must sign, are:

- to meet all regulations (i.e. compliance);
- to operate to industry best practice standards;
- to assess health, safety and environmental impacts, both actual and potential;
- to 'work closely with the authorities and the community in achieving the required levels of performance';
- to disclose 'relevant information' (CIA 1992a, p.4).

These principles are then detailed in guidance notes and codes of practice, which the CIA argues meet those for BS 7750, IS9000 series and (probably) EMAS (see Chapter 4).

As noted above, the two principal aims of self-regulation such as this are pre-empting regulation and increasing legitimation. The CIA supports legislation which is 'cost-effective and workable', 'realistic' and 'practical' and harmonised across European borders, but not 'excessive' (CIA 1992b, p.2). Legislation that meets these criteria would protect the reputation of the industry and its investments, but not penalise those who had already invested in environmental technology by advancing the regulatory requirements beyond their current best practice. In order to ensure this protection, the CIA lists its objectives in *The New Agenda* (1993, p.10) as:

- getting favourable legislation;
- encouraging voluntary action; and
- promoting best practice.

The commitment to voluntary initiatives such as Responsible Care is therefore driven by the aim of pre-empting legislation and regulation by presenting a workable industry-supported alternative, which is of course favourable to the industry:

> it is vital for the industry to maintain momentum in its voluntary initiatives in order to demonstrate the viability of its own proposals in anticipation of increasingly prescriptive legislation. (CIA 1993, p.10)

Part of this viability has been linked to the industry's claim to expertise, that its own experts are the most able to develop appropriate regulatory criteria on the basis of their specialist knowledge (Simmons and Wynne 1993, p.218). The argument runs thus: only industry has adequate experience of the technologies and economics of its operations and therefore only it can truly know how far these can be changed. If regulators do not take industrial advice, then they will set impossible standards far beyond current technological or economic capability, and therefore compliance will be poor.

> The fundamental difficulty for regulators is that they know less about the costs of complying with regulations than firms know themselves. One implication of this asymmetry of knowledge is that the regulator cannot simply tell firms what to do and be sure of the outcome . . . And as long as regulators need guidance from firms as to what the costs will be, business has a legitimate voice in shaping the law. (Barrett 1991, p.7)

Hence, this self-regulation is the pre-emptive response of an industry which sees more legislation as continually threatening and, in the long term, inevitable. However, if the self-regulation is credited as effective, it may be incorporated into the final form of the external regulation, thereby ensuring that business interests are appreciated. Although prompted by reaction, the approach is also attempting proaction in the regulatory agenda.

> The main regulatory issue for the CIA . . . is not one of replacing state regulation with self-regulation, but rather of negotiating a balance between the two that satisfies both the normative requirements of regulators and the operational preferences of its members. (Simmons and Wynne 1993, p.210)

Simmons and Wynne (1993) thus argue that Reponsible Care is a form of symbolic politics which claims industrial accountability to political pressure but only represents a 'small step' towards fuller cultural change. It certainly established a positive, self-regulatory alternative to the political push for external control.

However, the public credibility on which this rests is being tested. The CIA cannot force compliance with the Responsible Care criteria once companies have signed up to them. Like the Packaging Standards Council, it has no sanctions over its (voluntary) membership, nor does it undertake strict monitoring and information gathering on those signing up to Responsible Care. In fact, the CIA rejected such a strong policing role because it clashed with the traditional role of the association in supporting its members (Simmons and Wynne 1993, p.208). It seems that companies in the CIA network, not the CIA itself, conduct the only monitoring, through encouraging and checking on their competitors, prompted by the conscious linking of company self-interest to the improvement of the collective performance and image of the industry (Simmons and Wynne 1993, p.212). A good sectoral image will therefore rub off on companies' own individual reputations and economic viability as part of a publicly more legitimate industry.

The CIA restricts itself to 'encouraging' members to supply data on emissions, accidents, energy consumption and complaints, rather than forcing them into monitoring procedures. It then aggregates these statistics across the industry as evidence of the successes of Responsible Care (CIA 1992a, 1993, p.6). But it is difficult to see how such a voluntary system can fully discharge its environmental and social accountability to members as Gray *et al.* (1987) argue is important (Chapter 3). Unlike in regulatory systems, the pressures on companies to change are avoidable and there is little guarantee that a lack of change even in signatory companies will be identified and penalised. The consequent credibility problem has been recognised, albeit couched in the positive message of Responsible Care, by Dow:

> Public skeptics of Responsible Care may doubt the credibility of a self-policing industrial program relying on annual self-reporting. We think that seeing our continuing reductions of emissions and incidents will lead to believing. However, we also believe that we need to develop some means of independent validation of our progress if Responsible Care is to have ultimate credibility with the public. (Dow Chemical Company in Smart 1992, pp.71–2)

Hence, self-regulation has to be seen as credible in order to prevent government intervention, to reinforce the idea that industrial policing can be as effective in ensuring change as regulatory action. This reinforces the problems noted above with self-regulation as a proactive strategy. This has been highlighted by the variable patterns of environmental impact in aggregate statistics released by the CIA. For example, in a sample of 92% of its members in 1994, investment in environmental protection had fallen in real terms by 17% since 1993, whilst energy efficiency improved by 7.5% (*ENDS Report* 1995, 245, p.4). The latter change is beneficial for the businesses in cost terms, but the former is more disturbing in a sample of companies specifically pledged to environmentally responsible operations.

So, the possibilities of pre-empting regulation depend on the credibility of the self-regulatory scheme. This credibility is also important to business more widely. One of the explicit aims of Responsible Care is to pre-empt bad publicity (especially locally) in the light of the many widely reported chemical incidents in recent decades (e.g. the leak of gas from Union Carbide's chemical plant in 1984 in Bhopal, India). These proved both publicly and financially damaging to the sector as a whole because companies in similar fields were tarred with the same brush in public perceptions. Legitimation was consequently identified as critical to business survival, such that 'no industrial enterprise can survive and flourish in the long term without its operations being acceptable to the society within which it operates' (CIA 1992b, p.1). Such publicity problems needed to be addressed through Responsible Care, as the last paragraph of the Responsible Care booklet for the UK clearly stated:

> Gaining public confidence and trust depends on an open attitude to information and a co-ordinated management effort to maintain closer neighbourhood relations. This is the key to the industry retaining the licence to operate. (CIA 1992a, p.11; also CIA (Director General) 1993, p.2)

This also reinforces the first aim of pre-empting legislative and regulatory moves, whereby favourably influencing public views can help to relieve the pressure on governments to legislate (Simmons and Wynne 1993, p.205).

Evaluating environmental self-regulation

The examples of the PSC and Responsible Care suggest that the potential influence of environmental self-regulation is threefold. First, self-regulatory schemes may aim to make legislation redundant and therefore prevent it. This benefits both business, which gets criteria that it favours, and government, which has lower administration and costs and can also attempt to transfer legitimation problems to the self-regulators. However, the prevention of legislation is unlikely in the long term for environmental issues. The pressure to initiate, integrate and strengthen environmental regulation from the EC is clear and, despite the UK government's attempts to avoid EC regulation in other areas such as social policy and its own preference for deregulation, environmental regulation in the UK is continually expanding in response (Grant 1993, p.58). In contrast, in the short term, business may be able to delay or weaken legislative moves, especially where public opinion is appeased on some issues and where recession makes the impetus towards stricter regulation sluggish.

Secondly, self-regulation aims to increase the legitimation of business environmental practices and hence to commandeer more consistent political support. The public success of this aim is doubtful, especially in industries perceived to be highly polluting—which of course are the ones which have greater legitimation needs in the first place, such as the chemicals sector. Retailers and packaging manufacturers have yet to earn a strongly negative public image in relation to their environmental effects and they have therefore been less concerned about developing self-regulation. Their activities have been carried forward by companies, with associations playing a back-seat role (e.g. the B&Q intiative in Chapter 2). This is really a circular argument: where legitimation is most needed self-regulation develops, but it is also least likely to gain that legitimation. Even where it provides information and monitoring, which is rare, it does not necessarily close the 'legitimacy gap' (Sethi 1981) of public distrust because this stems not from a lack of information but from a more deeply perceived differential in the power and access to knowledge between public and business.

> As long as the Responsible Care program is founded on the premise that furnishing the public with evidence of its improvement will rebuild trust in the chemical industry, it is unlikely to achieve the recuperation of public confidence that the CIA is hoping for. (Simmons and Wynne 1993, p.219)

And, as noted in Chapter 2, the lack of public confidence will also rebound on political confidence, undermining the industry's attempt to pitch self-regulation as a legitimate replacement for governmental action.

Thirdly, and less importantly than the first two rationales, self-regulation can act as a focus for 'best practice' in particular fields, hence providing a blueprint for legislative moves if these become unstoppable under increased public pressure. This gives industry a positive breathing space to ensure consistent and consensual adoption of what the sector considers achievable improvements. The same circular argument applies here, with the qualification that, in cross-sectoral regulatory schemes, codes of practice might apply to a variety of businesses and lead to greater consensus across sectors, e.g. for environmental reporting and advertising (Chapter 4).

INFLUENCING AND LOBBYING ON ENVIRONMENTAL POLICY

As well as self-regulatory moves to pre-empt legislation and political pressure, business has developed more explicit lobbying strategies to influence policy and regulatory development. The success of such lobbying obviously depends on the political climate, including the influence of environmental NGOs on this (see Chapter 2), as well as on the type of regulation in question, its proposed remediation and the strength and credibility of the business argument. All these aspects vary from country to country and from issue to issue.

In the UK, the legislative climate for environmental issues has tended to be very discretionary, involving 'enabling' legislation which sets the parameters but allows the details to be worked out through consultation with business and other groups and permits subsequent changes in accordance with developments in technology and environmental science. This means that there is considerable latitude for business influence, as flexibility is built into the system. This contrasts with the situation in other countries where there is more prescriptive legislation, e.g. Germany and Denmark, or where the relationship between regulators and the regulated industries is less close, e.g. in the USA. The UK discretionary tradition thus facilitates legislative influence by business.

For example, the strong industrial lobby that developed regarding the implementation of Integrated Pollution Control (IPC) in the Environmental Protection Act (EPA) 1990 has been credited with a number of successes. The Environmental Protection Bill proposed to begin implementing IPC in January 1991 but the final Act postponed this to April 1991 for large combustion plants and new or greatly modified processes, and to April 1992 for all other plants and processes (existing chemical processes were granted a further 18 months' postponement in 1993). Considering that the process of implementation itself is scheduled to take four years, it will therefore be difficult to measure the success of the EPA 1990 fully until after April 1996. Even in the early stages, industry has continually appealed against the conditions set by HMIP for techniques to be accepted under IPC, e.g. ICI in 1992 against the specification of permitted chemical manufacturing processes (*ENDS Report* 1992, 213, p.19). It has argued for more certainty and detail in the application of IPC, as well as for a greater appreciation of industrial constraints when operationalising the concepts involved. Its sectoral association, the CIA, has sought to put forward an alternative methodology for adoption by HMIP when assessing 'best practicable environmental option' under IPC, a methodology which depends more on commercial questions than environmental effects and which would therefore push the entire regulatory process more towards business thinking (*ENDS Report* 1995, 249, p.25). Further, during the deregulation initiative in 1993–4, the UK government explicitly sought to help business. For example, where regulatory agencies were pushing for compliance with environmental regulation, they were required to detail the procedures involved and to notify the companies concerned of their intention to enforce such procedures, giving companies both warning and flexibility to facilitate their challenging of regulation (*ENDS Report* 1994, 237, p.27). Hence, whilst it is unlikely that business lobbying will entirely prevent environmental legislation increasing in response to pressures (see above and Chapter 2), there are possibilities for business to ensure that its viewpoint is incorporated into the detail of environmental regulations—if its lobbying is sufficiently proactive and well targeted.

Analysing business influence

Before considering specifically environmental lobbying, it is worth outlining the nature of business lobbying more generally. In most countries, and certainly in the

UK, business has a higher degree of access to policy makers than have NGOs. In practical terms, this is related to the greater resourcing of business compared to environmentalist NGOs and its closer ideological similarities with the current UK government. In theoretical terms, it is aided by the structural relations between government and business, where the economic power of the latter translates into political importance (e.g. Blowers 1984, p.214; Offe 1981, p.124). Corporatism is a key theoretical interpretation of how business groups influence policy and politics. Although the wider debates about the value of the corporatist model are tangential to the scope of this book,[2] corporatist theories introduce two mutually influential notions which are relevant here: negotiation (through 'intermediation') and implementation.

Negotiation is where state bodies discuss legislation with other groups, e.g. business associations, rather than merely dictate its final form directly to those groups. Hence, there is considerable bargaining between the two sides through specific channels and committees, both formally and informally. Implementation refers to the involvement of business groups in the fulfilment of legislation. In this way, business practice and specifically self-regulation reduce the involvement of government regulatory agencies, such as HMIP and the NRA, by taking on industrial policing, as suggested above. In such a case, business groups would become allies of government and would control their members in its interests, in return for having considerable influence on policy formulation. This may mean that government can gain greater influence through such cooperation, especially in policy areas where it might not be able to intervene so effectively alone. So business obtains policy influence and government obtains industrial influence, underlining the supposed mutual benefit of private interest government.

A corporatist model is made possible where the membership of a business association represents at least the majority of their sector, and ideally the whole. This enables the government to 'license' associations by recognising only one association per sector and thereby legitimating its existence and representation (Coleman and Grant 1988, pp.481–2). However, the 'bargain' in this licence assumes that associations will help the government ensure compliance with policy, e.g. through codes of conduct that are monitored across their memberships (Williamson 1989, p.214). So, the potential of associations to help in implementation should match the privilege and negotiation status accorded them by government. In this way, privileged policy influence is institutionalised and continually reinforced (Grant 1984, p.5).

A full corporatist model cannot be applied to many European countries, although researchers disagree as to which countries are closest or furthest away from this model in practice (see Williamson 1989; Keman and Pennings 1995; Crepaz and Lijphart 1995). Grant (1983, 1987) has concluded that the UK in the 1980s was generally a 'company state', rather than a corporatist one. However, self-regulation and other business–government agreements that fit a corporatist model may still occur in certain policy sectors (Grant 1992, pp.54-5), representing 'meso-corporatism'. The current UK Government's distaste for legislative intervention on principle, coupled with its distaste for EC legislation specifically, has encouraged voluntary agreements and self-regulation, e.g. of energy labelling of domestic electrical appliances, representing limited corporatist development (Sargeant 1985, p.250). It appears that some corporatist ideas might be useful in clarifying the business influence on environmental policy whilst acknowledging that the overall situation does not match the requirements of an entirely corporatist business–government relationship. 'The discovery of a variety of resilient meso corporatist arrangements might seem to be a less exciting conclusion than was promised by the original macro hypotheses formulated by corporatist theorists', but it might be more significant in understanding the influence of business in the 1990s (Grant 1992, p.56).

As well as theorising about the relations between business and government, it is important to appreciate the lobbying tactics brought to bear in the interests of business. The best time to influence legislation and policy development is at the earliest stage, when there is less entrenched opinion or commitment that has to be overturned (Jordan 1991b, p.180). Once policy is publicly announced by politicians, their credibility and legitimacy would be threatened by subsequent large policy shifts. However, whilst policy is still being drafted and sent out for consultation, commitment is less sure and business is more able to influence the eventual decision. Secondly, marginal changes to policy are easier to lobby for than overall reorientation. Incremental lobbying over a long period may bear more fruit than an out-and-out attack on a large section of policy which is highly politically significant. Until the late 1980s, when environmental issues became more politically sensitive and broader in policy scope, this was the business approach to environmental policy. As Grant (1983, p.79) noted during that earlier period:

> the CBI's tactics are undramatic and its success difficult to measure. Year by year, the CBI works quietly away on environmental legislation . . . endeavouring to modify the delegated legislation that often puts environmental measures into effect through detailed negotiations with the government department concerned. It is all very unspectacular, but it does bring results which benefit the CBI's members.

With the rising interest in environmental issues, business has more dramatic and wider influence on not merely the technical details of environmental regulation but the overall orientation and prioritisation of a rapidly expanding environmental agenda (e.g. Chapter 6).

There are also different scales of policy to be influenced. Groups such as the CBI (see later) and the ICC (see Chapter 6), which cover a variety of industrial sectors and company sizes, are more likely to look at wide-scope, long-term policy development, whereas sectoral associations tend to look at the detailed regulatory specifics. At all scales, the credibility of lobbying arguments is very important. This depends on maintaining trust between business and government representatives over long periods, often through staff continuity. The ideal is to make the policy official dependent on the industrial lobbyist for the specialist information that the latter provides (Jordan 1991b, p.180). The nature of the information is also highly important. Priority is placed on its being 'scientific' and 'sensible', not dramatic or unconventional as that provided by environmental NGOs may seem to be in order to gain media attention (Wilson 1990, p.68). Hence, the aim of business lobbyists is 'the undramatic submission of a well researched, well argued and representative case . . . in practice most lobbying is quiet and businesslike' (Miller, professional lobbyist, 1991, p.57). This is part of 'professionalising' lobbying by trying:

> to make the issue under discussion as apolitical as possible. When a policy is 'depoliticised' it can be resolved in industry/departmental negotiations more easily than when it is treated as a matter of political principle. (Jordan 1991b, p.185)

In terms of style, then, business associations tend to play down drama and moral arguments and rather seek to 'depoliticise' their case in order to enhance their credibility. Additionally, lobbying may draw on self-regulation promoted by associations in order to reinforce its legitimacy by publicising its existing but voluntary commitments to change.

In the 1970s, business lobbying concentrated on the civil service as being where the 'real power' lay (Miller 1991, p.49), leading to the professionalisation and expansion of lobbying (Jordan 1991a). However, by the 1980s, parliament was 'back in the frame' as an important but secondary target of lobbyists, including representation to Select Committees, MPs with constituency interests in specific industrial sectors and ministers with predispositions towards specific issues (Miller 1991, p.59; Jordan 1991b, p.173; 1991a, p.40). Currently, lobbying is directed at two sets of targets:

• elected representatives in both the UK parliament and government (MPs, peers and especially Ministers); and
• officials in government departments and the civil service generally, especially the DTI and the DOE on environmental issues.

In order to explore business environmental lobbying further, I shall consider the activities of some prominent players, including different business associations as well as more loosely organised initiatives and large companies. The influence of the last group can be more difficult to monitor for research purposes, forcing us to look more to the associations whose 'trail' of influence is discernible through publicly available publications and written policies, unlike the often one-to-one contacts maintained by executives in major companies with the government and civil service. Associations have also been expanding their environmental activities in recent years, giving considerable material for analysis. Unlike in Germany, UK associations are on average poorly representative of their sectors and peak associations (which cover many sectors and embrace other associations as members) are particularly under-representative, although they prefer to claim that they speak for the majority, at least of their sector, in order to boost their political clout. Further, often the activities of different business associations and groups on particular issues overlap or duplicate, suggesting that business representation has not yet established clear streamlining or effective coalitions in the UK in many policy areas (Coleman 1990; Grant 1987, p.173). This can cause fragmentation of business lobbying on an issue and prevent the collective representation of a case to government, e.g. on packaging as discussed later.

The Confederation of British Industry (CBI)

We shall begin with the CBI, probably the most publicly well-known business association in the UK. It was formed in 1965 from the merger of the Federation of British Industry, the British Employers Confederation and the National Association of British Manufacturers. By 1994, over 250 000 companies from a range of manufacturing and commercial sectors and over 200 associations were members and contributed to an annual income of over £17 million in 1993. It would like to be the overall business voice in the UK, but this is hampered by the fact that it is probably closer to manufacturing industry than to the service or financial sectors (Grant 1983, p.77) and, although 90% of its 1993 company members employed fewer than 200 employees, its central committees are dominated by representatives of much larger UK and multinational companies such as Shell, British Telecom, Guinness, Hanson and British Petroleum. It aims to be comprehensive in its sectoral and issue coverage (Coleman and Grant 1988, p.477) and its primary objective is:

> to voice the views of its members to ensure that governments of whatever political complexion—and society as a whole—understand both the needs of British business and the contribution it makes to the well-being of the nation. (CBI 1994, p.1)

A leaflet produced in 1990 reported results from a survey where respondents ranked various CBI roles according to their importance (Table 5.1). Although no percentages were given for the responses and the sample is vague, the rankings suggest that respondents see the CBI as more significant in influencing legislation and regulation than associations in general (which is no doubt why the CBI publicised the results of the survey). It also makes explicit the role of associations in being proactive, not merely defensive or passive in the face of changing environmental pressures. In order to strengthen this role, the CBI has consciously chosen a twofold approach to make a 'constructive contribution to attitudes and forward thinking' in the longer term but to react 'quickly and positively' in the shorter term to proposals from government or other sectors (CBI 1990, p.8). From comments made by other UK sectoral business associations in interviews with myself in 1994, it seems that they focus steadily on reacting to specific regulatory developments but consciously look to the CBI for a more strategic input to the wider debates.

The CBI's environmental policy unit has spread its attention over a variety of issues, including Integrated Pollution Control, contaminated land, water regulation (due to the importance of the newly privatised water companies in the 1990s) and

Table 5.1 Perceptions of the roles of the CBI and of trade associations generally

Rankings for CBI	Roles	Rankings for trade associations generally
1	Advising legislators	3=
2	Shaping regulation	5
3	Advising members	1
4	Promoting assessment	3=
5	Promoting good practice	2
6	Identifying costs	6

Source: adapted from results of members' survey in CBI 1990 (roles ranked in order of importance)

environmental reporting. One particular self-regulatory environmental initiative is the Environment Business Forum, launched by the CBI in February 1992 as part of 'an increasingly proactive stance by the CBI and its members' (CBI Environment Head in interview 1994). It was set up as a network for companies prepared to make an environmental commitment and improve their environmental performance via written action plans, environmental targets and public reporting of progress. The Forum was primarily to provide guidance on environmental management both from the CBI to companies and from company to company through networking, but also to show that 'business-led voluntary action will be more effective for the long-term protection of the environment than legislation alone' (CBI Annual Report 1993, p.23). It therefore has an internal function within business in promoting good practice and an external and legitimating function in representing business change to other bodies, especially to government, similar to other self-regulatory schemes noted earlier in this chapter.

On joining the Forum, companies were to draw up specific targets under an 'agenda for voluntary action' and later they were to report on their progress, because 'only by reporting openly on corporate environmental performance in this way can business ensure its long term viability' (CBI *Environment Newsletter* 1993, 10, p.3). This echoes comments made by the CIA about its self-regulatory Responsible Care programme above. The CBI claimed that many benefits would come from companies joining the Environment Business Forum (Table 5.2) which again prioritised the role of self-regulation as pre-empting national and European legislation, but also pointed to the operational benefits of improved environmental performance.

However, the Forum ran into difficulties. The original target set was for 1000 companies to be registered by 1994, but only about 250 had signed up by the middle of that year. In addition, *ENDS Report* (1994, 230, p.3) suggested that about 50 of these were not even individual companies but local authorities, consultants and NGOs, further eroding the total. Many companies were not producing the required environmental reports, although the CBI was by 1993

Table 5.2   Suggested benefits from joining the CBI's Environment Business Forum

---

An opportunity to influence UK and EC policies on environmental issues by demonstrating the effectiveness of business-led voluntary action.

Participating in a network of organisations committed to the goal of environmental excellence, allowing organisations to meet and exchange ideas, information and experiences.

Access to high quality advice and up-to-date information and guidance on environmental issues from the business perspective.

Competitive advantage to business through involvement in a major initiative to enhance the environmental performance of Forum members and of the UK in general.

Source: adapted from CBI leaflet *Voluntary Action Counts* (n.d.)

accepting in their stead 'reasons for delay with commitment to a revised timetable' (CBI Annual Report 1993, p.23). The disappointing performance of the Forum so far has dented its credibility and weakened the CBI's argument that voluntary environmental initiatives would be more successfully implemented by businesses than would external regulations or taxes (*ENDS Report* 1994, 230, p.3). It is perhaps the requirement to publicise progress that is deterring companies from joining the Forum because this is rarely seen as beneficial by business (see Chapter 1). This has been recognised by the CBI:

> The actual process of setting targets and reporting against them is one which businesses are uncomfortable with, extremely uncomfortable with, for a number of different reasons. And from our point of view as an association, it's frustrating, although I understand why: companies are doing a lot more than they want to talk about. (CBI Environment Head in interview 1994)

So, the CBI's positive programme faltered and, in a major report in 1994, it changed tack by lobbying government to pay more attention to 'environmental costs' and calling for a stronger debate about the effects of environmental and health and safety legislation on competitiveness, and for a more effective business contribution to that debate. Specifically, the CBI called for government to provide clearer, more streamlined regulation and to defend business against the application of more European regulation. They also called for business to develop better environmental strategies and to work on their environmental and health and safety 'cultures' (see Chapter 3). The CBI's environmental stance therefore seems to be becoming more overtly defensive than the Forum set out to be. This reflects not only the economic constraints of the continued recession but also the use of financial arguments against the moral and political ones of environmentalist NGOs. The CBI is increasingly concerned to protect the status quo in the face of expanding environmental legislation from all sides, an encroachment which was not halted by the government's movements towards deregulation (Grant 1993, p.58).

Generally, the CBI supports voluntary action as a first response to environmental issues and only supports economic incentives and regulations as secondary responses where these do not adversely affect business performance (e.g. CBI Annual Report 1993, p.24). However, voluntary action is complicated by the CBI's varied membership, which works against the development of consensus because different sectors will support different actions. The responsibilities of those sectoral associations which are CBI members are often confused, making it difficult for the CBI as a peak association to coordinate activities.

> Whom the association is representing becomes less clear and the ability to coerce or persuade 'members' to follow policy agreements declines. (Coleman 1990, p.247)

This is particularly true for cross-sectoral issues where the range of viewpoints to be accommodated is large. This can be illustrated with reference to packaging:

> one of the criticisms that is often, that is *sometimes* made of the CBI is that it can become lowest common denominator so a great deal of effort goes into *not* being that. The area where we have steered clear, and because there are so many subjects we can deal with I suppose in a sense it's sometimes a relief *not* to be able to intervene, is packaging and recycling which is *unbelievably* difficult. (Environment Unit head in interview 1994)

So the CBI has focused on strategic environmental issues rather than those which tend to be sectorally divisive. For example, it produced *Firm Foundations* in 1993 to argue on behalf of all sectors that liability for contaminated land should not be strengthened but merely clarified and set within the standards of industrial best practice (CBI Annual Report 1993, p.22; cf. the US situation, Chapter 2). This emphasised the problems for the CBI's policy staff in developing a viewpoint which could be useful in clearly informing policy makers but which did not differentiate between sectors.

> Ultimately, the CBI is governed by consensus, but it is a consensus which rests on an interpretation of what the membership, particularly the more influential of them, want, rather than on any clear statement of policy preferences by them. (Grant 1983, p.83)

The Institute of Directors (IOD)

Although the CBI is an important representative of British business (Grant and Marsh 1977), its influence is contested by other comprehensive business associations which seek to put forward 'the' business viewpoint, one being the Institute of Directors (IOD). The IOD was established in 1903 and all its members are individual company directors, or people of similar position, not companies or associations. It had about 34 000 members in the UK in 1993, about 60% of these being directors of SMEs, and a further 14 000 or so members in other countries.

As well as a lobbying organisation, it presents its Pall Mall facilities as 'London's premier business club'. Its lobbying function is bound up with economic growth:

> our object is to ensure that governments establish an economic and regulatory environment in which our members' businesses can flourish, expand and compete internationally. (IOD brochure, n.d.)

Like the CBI, the IOD identifies a more strategic role for its input which is not restricted to the technical details of policy and regulation but also looks to changing 'the climate of ideas . . . a sort of think-tank role' (IOD environmental policy coordinator in interview 1994). However, unlike other sectoral associations, e.g. the CIA and even the CBI, it is not dominated by large companies. Its concentration on the SME viewpoint means that it particularly advocates free trade and less regulation, whereas other associations may support specific regulatory initiatives where these protect the investments of their members (the National Association of Waste Disposal Contractors did so over waste management licensing in the UK in 1994). The IOD may be closer in philosophy to the 'new right' political ideologies of the UK government in the 1980s and 1990s than is the more staid CBI. Coleman and Grant (1988, p.478) argued that this closeness helped it to contest the CBI's role as the sole business voice in the 1980s, so that 'the star of the Institute of Directors, with its more conservative viewpoint, has brightened and that of the CBI has dimmed' (Coleman 1990, p.247).

In environmental terms, however, the IOD was 'invisible' until 1992 when it began producing publications relating to environmental issues (*ENDS Report* 1993, 227, p.24). This perhaps reflects the slow response of SMEs to environmental issues, as noted in previous chapters. They tend to be more compliance-oriented than do very large companies, because the costs of technologies to ensure compliance constitute a much higher proportion of their total costs.

Then, in 1992, the IOD published a statement on environmental issues as *Stewards of the Earth: Environmental Policy for a Market Economy*, which spent 14 pages outlining government responsibilities on the environment and only 4 pages on outlining industrial responsibilities. This report cast regulation as the 'last resort' for environmental protection (IOD 1992, p.4) because it might act as a barrier to industry and distort its markets. Further, and against the thrust of the precautionary principle, the IOD wanted to dissuade the government from taking what the IOD regarded as precipitate regulatory action on emerging environmental problems: 'We would recommend against the slavish following of each new environmental fad that emerges' (IOD 1992, p.15).

So, the IOD was seeking to avoid any further compliance burdens for its members in the form of new legislation, through some proactive lobbying. It was attempting, like the CBI, to focus on voluntary action as the first response to environmental issues, through self-regulation and self-policing of companies and the development of 'stewardship' policies, based on private property rights. However, these were general statements of policy, rather than detailed self-regulatory practices.

The IOD's philosophy is that private ownership of resources gives not only the right to their use, but also the responsibility for their stewardship. (IOD 1992, p.15)

In comparison, the IOD regarded economic instruments for environmental protection as less preferable and it rejected the carbon tax as a distortion of the market (as did many other business groups, see Chapter 6).

Since 1991, the IOD has commissioned an annual environmental survey of its members, asking questions about board time spent on environmental issues, the importance of those issues and company environmental policies (e.g. IOD 1993; 1995). However, the results from its members were not very encouraging about take-up of the environmental practices and policies which the IOD advocated in 1992. The proportion of respondents claiming to have formal environmental corporate policies was consistently around one-quarter across the four surveys between 1991 and 1994 (IOD 1995, p.8). Although the IOD argued that the recession was restricting such developments, it is clear that its arguments for voluntary action as a 'first response' to environmental issues were not being followed and that its members were still compliance-oriented. 'The credibility of the IOD's demand for voluntary action by business to be given priority is being undermined by trends within business' and within its own membership (*ENDS Report* 1993, 227, p.24). Moreover, its 1993 Annual Report did not mention environmental issues at all, despite the ongoing survey and the burgeoning environmental legislation that was bothering many other associations at this time.

So, although the IOD may find it easier to develop a consensus amongst its individual membership, unswayed by the differential powers of sectoral associations and large companies, its advocation of the status quo and of corporate environmental policies does not seem to have had a strong influence on the environmental agenda, nor has it yet presented any new initiatives in this area.

The Advisory Committee on Business and the Environment (ACBE)

It is interesting to compare the CBI and the IOD with a new business environmental initiative of the 1990s. The first Advisory Committee on Business and the Environment (ACBE) was set up in 1991 by the Department of the Environment (DOE) and the Department of Trade and Industry (DTI) specifically to enhance dialogue between government and business on environmental policy, as promised in the 1990 white paper (DOE 1990). It is not an association but a group of nominated individuals from major companies, and it has published progress reports with the help of specialist working groups, e.g. on the financial sector. The specific aims of ACBE are:

1. to conduct strategic level dialogue about long- and short-term environmental issues between government and business;

2. to mobilise business on good environmental practice and management (in cooperation with other organisations);
3. to be a focus for international business environmental initiatives (ACBE 1991).

With a secretariat jointly from the DOE and DTI, ACBE's role is to bring together thinking and offer recommendations to the departments, and it is made up of around 25 'business leaders' who serve for two years. The first ACBE was chaired by John Collins of Shell UK and lasted from 1991 to the end of 1993, when it was reconstituted for another two years to 1995 under Derek Wanless of National Westminster Bank. It is noticeable that many of the names of those on the two committees are familiar from other environmental business initiatives, and ACBE itself is pleased to relate its members' involvement in WICE (ICC environmental arm, see Chapters 4 and 6), the Business Council for Sustainable Development (see Chapter 6), Business in the Environment and the development of BS 7750 (ACBE 1992; see Chapter 4). Further, the individual members of ACBE obviously see their companies as taking the role of 'leaders' in environmental practice and they often reiterate pledges to demonstrate such symbolic 'top-level' commitment through their own operations as well as through ACBE's recommendations.

So what is ACBE's contribution to the government's environmental thinking? We can approach this by analysing the recommendations of the ACBE working groups. From the first, the Global Warming Working Group (ACBE 1991) supported energy efficiency, and thereby the Energy Efficiency Office (EEO) of the DOE, through energy labelling and rating schemes for buildings and products. This led to a *Practical Energy Saving Guide for Smaller Businesses: Save Money and Help the Environment* (EEO/ACBE 1993) to help SMEs monitor and manage their energy use, thereby saving costs. It also supported transport efficiency, on the basis that 70% of carbon dioxide emissions comes from vehicles (ACBE 1992, p.31). Its recommendations included the rejection of favourable tax policies for company cars, of high mileages and of large-engined vehicles, and support for financial incentives for efficient vehicles, diesel vehicles and market certainty to encourage investment in both.

This first set of recommendations was perhaps the most surprising, long-term and structural yet to emerge from ACBE. They included 'requiring businesses in major towns to pay a levy from their payroll towards public transport for employees', as well as road pricing, tax incentives to discourage car use and support for public tranport (ACBE 1992, pp.9–10). They also called for the Society of Motor Manufacturers and Traders in the UK (SMMT) to improve on its voluntary target of increasing fuel efficiency by 10% by 2005—which it did not (ACBE 1993b, p.43). Such ideas represent a fundamental critique of the current government's predilection for private transport and demonstrate ACBE's broad perspective on environmental change (especially considering that there were car manufacturers on the committee at this time). However, ACBE continued to focus also on the practicalities of company change and its consequent benefits, as we have noted business does throughout this book, producing *A Guide to Environmental Best Practice for Company Transport* (ACBE/DOE/DTI 1993). This stressed the cost savings to be made from environmental changes and encouraged the adoption of fuel-efficient vehicles, fuel consumption reduction targets (voluntarily set within companies), monitoring mechanisms and car-use management, including looking at employees' travel to work and commercial distribution practices. Other working groups looked further ahead than the current policy agenda. The Recycling Working Group supported landfill regulation and a landfill levy when the government did not (although in 1994 it was finally accepted by government and included in the upcoming legislation in opposition to work done by the Producer Responsibility Group, see later). The group also supported the use of recycled materials, to be encouraged through product standards and labelling, possibly through legislative controls.

But other working groups were less radical. The Environmental Management Working Group supported BS 7750 (Chapter 4), but only for minimal disclosure requirements and voluntary adoption, and the involvement of SMEs in environmental management. This group therefore supported existing policy and developments with a more predictable business attitude. Similarly, the Commercial and Export Opportunities Working Group (set up in 1992) advocated more information provision about commercial opportunities but little in the way of policy initiatives or changes. The Financial Sector Working Group (set up in 1992) especially concentrated on the issues surrounding liability (see Chapter 2) but denounced any changes to the status quo—it merely called on government to confirm that there would be no changes, in order to quell uncertainty in the insurance market (ACBE 1993a, p.7).

ACBE was reconstituted in November 1993 with four new working groups and the re-establishment of the Financial Sector Working Group. This group continued to express the consensus against changes to liability, but also recommended encouragement for environmental reporting, to which the government responded that responsibility should remain with the financial sector (ACBE 1994, pp.14–16). There was therefore little government support for new developments. The Achieving Environmental Goals Working Group clearly rejected further regulation by advocating that government take 'the least prescriptive and onerous means of achieving environmental objectives' (ACBE 1994, p.19) and felt that the government agreed. It also called for economic justification of environmental policy proposals, to make policy link with 'its scientific and economic merits' not with 'uninformed public concern' (ACBE 1994, p.23) and to develop a strategic and generic approach to all environmental issues based on 'value for money' (ACBE 1994, p.20). The group produced a 'test kit' of questions which it considered should be asked of any proposal before government gave its support—the questions emphasise cost, competitiveness, public response and criteria for success rather than environmental implications (ACBE 1994, p.22). The Waste Minimisation Working Group was influenced by the importance of packaging waste (see later) and again emphasised that any policy should address cost-effectiveness and implement economic incentives (e.g. a landfill levy) not legislation. They also suggested renaming waste minimisation 'Total Process Efficiency' (TPE) in the belief that this positive wording would encourage its adoption, a suggestion which the government (sensibly) rejected. Even less radically, the Transport Working Group diluted the strong suggestions from the first ACBE by advocating research into alternative fuels and tentatively supporting road distribution rather than a movement towards rail freight. Overall, the emphasis on public transport was lost and the strength and willingness to propose changes to the status quo dissipated.

Government response to ACBE has been variable. DOE and DTI ministers must respond to each progress report, but they 'passed' the recommendations made in the first ACBE for fiscal and policy measures against cars to the Treasury instead of responding through their own departments (ACBE 1992, p.13), or described them as 'commercial' issues for operators (ACBE 1992, p.20). There was, in other words, little support for ACBE's suggested measures.

> The Government is keeping under review whether further measures may be necessary to discourage car use by companies in congested areas. (ACBE 1992, p.19)

The first ACBE was 'disappointed' with government responses to its suggestions (ACBE 1993b, p.14) and this is reflected in the weaker, shorter-term recommendations from the second committee, which are generally more in line with current government policy.[3]

So ACBE was set up to be a voice of consensus on business and the environment, uncontaminated by sectoral arguments. The government considered that this was not already provided by existing associations such as the CBI, which represented a diversity of attitudes across the breadth of business. Indeed, the Financial Sector Working Group noted that its first report was 'the first time that the financial sector has formally and jointly considered its response to the environmental agenda' (ACBE 1993a, p.5). At the conclusion of the first committee, its outgoing chair said that ACBE 'had received many plaudits, both domestically and internationally' (ACBE 1993b, p.4) and, at the conclusion of its second, that it was 'highly regarded by business' and that the government had been 'equally enthusiastic' (ACBE 1996, p.4). Yet, it is difficult to see clear traces of its influence on policy. Partly this is linked to the need for government support, funding and legislative innovation, which both incarnations of ACBE stressed but neither received. They continued to argue that government should take a 'share of the responsibility' for environmental actions (ACBE 1993b, p.6) without much evidence of this happening. It is also linked to the way in which the reports frequently endorsed the status quo, rather than seeking regulatory modification. Where a more proactive stance was adopted, this was usually with reference to innovative operations or best practice rather than to environmental policies. We could therefore argue that ACBE was an attempt by government to 'organise' environmental issues out of mainstream politics, as noted above. Alternatively, ACBE was seen by business as a chance to influence policy on key issues. In neither case does it seem to have had a huge impact on the orientation of either the business or the environmental agenda.

Companies

> Large companies can, in their own interests, be as active and effective as pressure groups or trade associations. (Jordan 1991a, p.16)

As well as associations and initiatives seeking to influence the environmental agenda on behalf of business interests, environmental lobbying can be performed directly by companies. Although operational change in companies has been widely investigated in the literature, their lobbying influence has been less studied. This is because the influence of associations is easier to trace directly, being more publicised and collective.

However, the political input of larger companies to the environmental agenda might be traced in two forms. First, they hire commercial lobbying companies on their behalf or develop in-house government relation departments, e.g. in Shell, British Petroleum and ICI (Grant et al. 1989; Grant 1987, p.171). These tend to develop specific lobbying in an *ad hoc* manner on issues of key concern to those companies by organising meetings with MPs and civil servants. Secondly, individuals, again mainly from large companies but also from companies considered to be environmentally aware or 'leading', join together under the umbrella of policy discussion committees. Obviously, secondment of staff or other forms of company contribution to associations, consultation committees and similar groups ensures company representation through individuals in key business–government fora. For example, the UK Ecolabelling Board in 1994 had amongst its members Derek Norman (Pilkington plc, the glass manufacturer), Nigel Whittaker (Kingfisher plc), Keith Humphreys (Rhône-Poulenc Ltd) and Jyoti Munsiff (Shell Transport and Trading). Although the Board is careful to declare that they are members 'in a personal capacity' (*Ecolabel Criteria* 1994, 7, December, p.6), this demonstrates the involvement of big companies through individuals. Grant et al. (1989, p.73) argued that the British government prefers to deal with companies rather than associations, and ministers may often rely on their close contacts with key individuals in those major companies. Useem (1984) has further argued that key individuals form an 'inner circle' or politically active group of top managers of large companies which drives the business lobby. He argues that the power of this circle transcends national boundaries, sectoral differences and political divisions in affecting government policy and public opinion. Thus, the viewpoint of the large companies is diffused through individual contacts by its top people to all levels of government. Again, this can be difficult to trace and has as yet not been satisfactorily explored in the business-environment literature.[4] However, the establishment of new and publicly advertised environmental committees, e.g. ACBE (see above) and PRG (see below), suggests a governmental desire for a more collective input to policy discussions.

As well as acting individually, large companies are also able to dominate associations in particular sectors, e.g. chemicals, energy, water services, where they are the main players and can mould association policies to their interests. Even in comprehensive associations such as the CBI, large companies are key actors in policy formulation (Grant 1984, p.7; 1983, p.74). Such domination is less clear where associations include a majority of SMEs or represent sectors where there is less monopoly, e.g. food and drink, plastics. In both cases, the influence of large companies tends to be diluted.

Of course, the 'behind the scenes' lobbying impact of large companies is reinforced by their more public image-building activities. Again, we can note how the public and governmental legitimation of companies is intertwined and reflect that the advertising, company policy and reporting material already discussed in Chapters 3 and 4 will inevitably also influence the environmental agenda and the political climate in which it develops.

BUSINESS AND PACKAGING REGULATION IN THE UK

I want to use the case of packaging and packaging waste regulation to illustrate the influence of business environmental lobbying in the UK. Packaging was a key cross-sectoral issue in 1994, involving manufacturers, retailers and raw material producers, as well as specialist producers of packaging (known as 'converters'). It involved a variety of environmental issues, from production of packaging and use of raw materials to consumption levels, generation of waste, disposal to landfill, recycling systems, incineration and composting. The case study will illustrate how business sought to influence a specific element of the national environmental agenda.

Activity in the UK arose due to pressures from Europe as individual countries developed their own legislation and the EC had to decide how best to harmonise these. The most well-known national legislation was the Packaging Ordinance passed by Germany in 1991, in force from January 1993, requiring the development of collection and recycling of packaging materials. Industry responded by setting up Duales System Deutschland (DSD) to organise the required collection of sales packaging (i.e. post-consumer) in a parallel operation to normal waste collection. This was funded by a retail levy on items marked with a 'green dot' (*grüne Punkt*). Business in the UK has criticised the way this scheme addressed only collection, ignoring the development of end-use markets for the collected material. Excess 'recyclate' (material collected for recycling) was therefore collected and much went to export in the early stages, affecting international recyclate markets and prices. Business characterised it as effectively a subsidy on recyclate, leading to a 'quasi-monopoly' in the recycling industry (*Financial Times* 1994, 16 March, p.18). A similar scheme was developed in France under its packaging decree in 1992 (effective in 1993). This was less strict than Germany's and counted energy recovery (i.e. waste being burnt or composted where energy produced is collected and used) as well as recycling as a way of dealing with packaging waste.

In an attempt to harmonise such national developments, the EC produced a draft Directive on packaging and packaging waste in 1993 which set targets of 60% of packaging waste to be recovered, including 40% specifically recycled, by 2000 and 90% and 60% respectively within 10 years of implementation. The Directive was based on a 'waste management hierarchy' as follows, in decreasing order of priority:

- prevention/minimisation;
- recovery/recycling (as equally valid);
- disposal/landfill.

The UK government had already, in its 1990 white paper, set a target of recycling 25% of domestic waste by 2000, but it had not clarified how this would be done.

The Consortium of the Packaging Chain (COPAC)

In response to the Directive and to increasing market difficulties ascribed to the German Ordinance, the UK government urged the Industry Council on Packaging and the Environment (INCPEN) to act and it helped to set up the Consortium of the Packaging Chain (COPAC). COPAC had three main aims:

- to put the business point of view into the discussions as early as possible;
- to develop a strategic UK response to European debates;
- and to meet the target identified by the government in its white paper of recycling 50% of recyclable domestic waste by 2000 (DOE 1990).

COPAC comprised six main business associations and a forum of associations, pulling in representation from all sectors of packaging manufacture, filling, distribution and retailing (Table 5.3). COPAC's final plan[5] set targets of recycling 33% of household waste across four sectors (glass, metals, plastics, paper) by 1999 (less than 10% of domestic waste). This was based on half of UK households being covered by collection systems and packaging only increasing by 1.5% per annum over that period.

COPAC tried to influence the agenda by reorienting the debate away from recycling to other operational options. For example, it extended the original concept of 'recycling' implied in the white paper by focusing on the 'valorisation of waste', defined as realising or enhancing the value of waste, i.e. recycling, reuse, composting and energy recovery, the last usually by incineration with electricity as a by-product (COPAC 1992a, p.3). This allowed certain sectoral positions to be supported: for example, plastics wanted a commitment to energy recovery without penalty to protect those industries which faced technical difficulties in recycling. In contrast, the established domestic recycling schemes of paper and glass through 'bring' systems (i.e. bottle banks) meant that these sectors supported recycling plans, not recovery targets. COPAC also urged that landfill be accepted as the 'best environmental option' for some areas and that packaging 'optimisation' (i.e.

Table 5.3 Membership of COPAC

Member association	Representation
British Retail Consortium (BRC)	Retailers
Food and Drink Federation (FDF)	45 associations of manufacturers
Industry Council for Packaging and the Environment (INCPEN)	Converters, manufacturers, retailers
Institute of Grocery Distribution (IGD)	Manufacturers, distributors, retailers
Packaging Manufacturers Forum (PMF)	Timber Packaging and Pallet Consortium (TPPC)
	Metal Packaging Manufacturers Association (MPMA)
	British Plastics Federation (BPF)
	British Glass (BG)
	British Carton Association (BCA)
	British Fibreboard Packaging Association (BFPA)
Cellulose Fibre Industry Group (CFIG)	British Paper and Board Industry Federation (now the Paper Federation) (BPBIF)
	BFPA
	BCA
	British Waste Paper Association (BWPA)
	Independent Waste Paper Processors Association (IWPPA)
Metal Packaging Materials Producers (MPMP)	Aluminium and steel raw materials producers

Source: adapted from COPAC (1992a, Annex 4). As well as the members listed, COPAC was also supported by technical advice from the European Recovery and Recycling Association (ERRA), the National Association of Waste Disposal Contractors (NAWDC) and the Organic Reclamation and Composting Association (ORCA).

minimisation) be emphasised, a position typical of INCPEN. Thus, COPAC was trying to redefine the waste management agenda in order to be both:

- positive in terms of commitment to a recycling target; and
- beneficial to business sectors where recycling is economically, technologically or ideologically constrained.

Hence, it favoured landfill, recovery and minimisation.

Coupled to this redefinition of waste management options, COPAC also urged the government to accept three things. First, it wanted responsibility for waste to be shared amongst government, consumers and industry, not be placed solely on industry's shoulders, because packaging was only a small proportion of the waste stream (c. 7% landfill waste, COPAC 1992a). Secondly, it wanted justification of proposed legislation to be based on 'sound environmental, technical and economic data' and some consideration of how to develop end-use markets (COPAC 1992a). This reflects arguments on legislative 'testing' by ACBE and IOD discussed above and is a very common business attitude. Thirdly and most importantly, COPAC demanded government support as a condition of further business involvement in recycling plans, especially support of UK business against the adoption of a German-style model in the EC Directive.

> COPAC is committed to assist in achieving the government's target and calls upon government to act in partnership with COPAC. (COPAC 1992a, p.6)

However, COPAC clearly implied that all costs for its Action Plan would be borne by industry (COPAC 1992b, p.4).

So, even as the EC was drafting its Directive, business was attempting to put its viewpoint into the debate through the circulation of COPAC publicity and its discussions with UK government. But the COPAC plan did not match the views of the new Secretary of State for the Environment, John Gummer. He took up the initiative in mid 1993, chose to bypass COPAC and set up a new body to address these issues.

The Producer Responsibility Industry Group (PRG)

In July 1993, John Gummer 'challenged' (i.e. instructed) industry to set up a Producer Responsibility Industry Group (PRG) to develop a plan by October of that year to address the recycling of packaging waste in the UK, under the palpable influence of European developments. The Environment Secretary wanted the PRG to put together a plan to recover between 50 and 75% of all packaging waste by 2000, a range similar to that in the draft Directive.

He invited 28 companies to form the PRG Steering Group (Table 5.4) who were mainly large, well-known and publicly acceptable companies. Nicknamed 'the great and the good' by representatives of other sectors which were not invited into the PRG, they were predominantly grocery superstore chains (ASDA Group, Argyll Group, which owns the Safeway chain, Tesco Stores, J Sainsbury) and food and drink businesses (Allied Lyons, Unigate, United Biscuits, Weetabix, Guinness, Bass, Coca-Cola Great Britain and Ireland) rather than the producers of packaging or its raw materials. The latter felt that it was they, not the retailers, who had the expertise in packaging and, equally importantly, that it was on their shoulders that the burden would fall under regulation. They therefore felt excluded by the ministerial selection process and perceived the PRG to be unbalanced and unrepresentative (unlike COPAC). Moreover, the PRG Steering Group was seen as repeating the data-gathering and disputation that COPAC had already dealt with, and yet having little in-depth knowledge of the waste and recycling issue.

The PRG's aim was 'to achieve environmental and economic benefits by recovering value from waste packaging materials' (PRG 1994a, p.2). It counted recycling, composting and energy recovery as equally valid options for waste. It set a target of 58% of packaging waste value recovery by 2000, defined as the amount

Table 5.4   Companies invited to form the PRG

Allied Lyons	Argyll Group
ASDA Group	BP Chemicals
Bass	The Boots Company
Booker	Coca-Cola Great Britain and Ireland
Bowater	Forte
Burton Group	ICI
Guinness	Kingfisher
Grand Metropolitan	Nestlé UK
Marks & Spencer	Procter & Gamble
Northern Foods	J Sainsbury
RHM	D S Smith (Holdings)
Shell Chemicals	Unigate
Tesco Stores	United Biscuits
Unilever	Weetabix

Source: PRG 1994, p.31

'diverted from landfill' (PRG 1994a, p.4) and made up of 44% of domestic waste and 70% of industrial waste (compared with COPAC in Table 5.5).

John Gummer had selected specific companies to design packaging waste self-regulation, suggesting that he saw the UK as a 'company state' (Grant 1993, p.13; 1987) and that he would rather deal with companies than associations for policy input. However, the PRG's original plan was equally, if not more, unsatisfactory for the government than COPAC's developments, not least because the PRG was set up in the full glare of media promise and at the Secretary of State's own initiative. The draft report, *Real Value from Packaging Waste*, was two months late and many of the details were qualified or left vague, a characteristic criticised by the House of Commons Environment Committee in 1994 and which the PRG blamed on time constraints (PRG 1994, p.1).

But it had also failed to decide where in the packaging chain to place the levy to raise recycling funds, even after calling in an 'independent expert' in the shape of Sir Sydney Lipworth QC to assess the question after wide consultation (Lipworth 1994). The obvious problem for the PRG in setting the packaging levy was that it would fall disproportionately on some industrial sectors. Hence, the PRG supported the notion of 'spreading the burden' of costs and administration through the packaging chain, couched in the language of 'maximising participation' (PRG 1994a, p.7). As Lipworth noted, 'individual companies or organisations clearly preferred the levy to be collected at some point in the chain other than their own' (Lipworth 1994, p.29). The PRG's problem was similar to the CBI's noted earlier. The diversity of sectors involved often brought arguments down to the lowest common denominator and frustrated the building of a 'unified business voice' to establish a desirable policy framework.

So the plan was left vague—business had not risen to John Gummer's 'challenge'

Table 5.5   Comparing PRG and COPAC

PRG Plan 1994	COPAC Action Plan 1992
Overall target: 58% recovered by 2000 (of domestic and industrial packaging waste) (50% recycled by 2000)[6]	42% recycled by 1999 (of glass, paper, plastic, metal domestic and industrial packaging waste)
Domestic target: 44% domestic packaging waste recovered by 2000 (35% recycled by 2000)[7]	Domestic target: 33% domestic packaging waste (glass, paper, plastic, metal) recycled by 1999
Mechanism: through VALPAK	Mechanism: normal business operations
Cost: £40 million per annum (£100 million in first draft) via VALPAK-administered levy for first few years	Cost: £50–55 million per annum via companies and FDF/IGD expenditure
Members are companies (chosen by the Secretary of State)	Members are associations

Sources: PRG (1994a); COPAC (1992a,b)

because there was no pre-existing coordinated business lobby on this issue, nor was it easy to establish one. The consequences of this were first that the PRG was not proactive in developing a business viewpoint and exerting lobbying influence through coordinated activity. Rather, it seems that sectoral associations and groupings lobbied the PRG itself for their own views to be heard and passed on through the new consultative channel. The cross-sectoral differences were not resolved and the message was weakened. Fortunately for business, the government was in any case reluctant to regulate even when the PRG was clearly having problems—the 'stick' of external control was not being waved with any force. In the end, like COPAC, the PRG essentially threw the ball back to government and called for legislative support for the plan but in more emphatic terms than COPAC had.

> Experience of earlier attempts to develop packaging recovery voluntarily suggests that some players would opt out from any purely voluntary involvement. This would jeopardise its accomplishment and place an undue burden on the conscientious participants . . . The plan can only move ahead with a commitment by government to provide the legislative backing to enforce compliance by all members of the packaging chain. (PRG 1994a, p.4)

The PRG therefore rejected voluntary self-regulation on packaging, despite the arguments cited in its favour earlier in this chapter. Hampered by a lack of consensus, it turned to calling for legislative backing as essential to not only success but any further cooperation by business in packaging policy.

We will tell [the government] that we can deliver the plan—if they deliver the legislation. (David Harding of PRG Steering Group, quoted in *ENDS Report* 1994, 236, p.13)

The PRG was disbanded after producing its plan in November 1994 and reincarnated as VALPAK—Working Representative Advisory Group (V-WRAG), which would set up the packaging scheme in more detail. The delays since John Gummer's challenge of July 1993 then allowed V-WRAG to argue that the targets for recycling increases by 2000 were now too rapid, implying a further slowdown in activity.

The business influence on government with respect to packaging policy has been to reinforce the status quo. Committees in the House of Lords (1993) and House of Commons (1994) reflected business views when they rejected EC legislation in principle as inappropriate for recycling and packaging waste. In its response, the government concurred, whilst lauding its own input to the current EC draft and its success 'in negotiating a Directive with a flexible, permissive nature, allowing member states to set targets in the most cost-effective way, in the light of national or local circumstances' (DOE/DTI 1994, p.11). However, on the problems of self-regulation, poor compliance and reduced competitiveness, the government finally seemed to succumb to industry arguments. In October 1994, it announced that it would legislate 'at the earliest opportunity' on producer responsibility for packaging waste such that:

> business initiatives on the recovery and recycling of waste are not undermined by those seeking to avoid any obligation to take part. It will be the minimum required to provide an effective deterrent to 'free-riders', and will preserve the ethos of the cost-effective industry-led approach. (DOE/DTI 1994, p.17)

So the government clearly held similar opinions to the PRG on many issues; for example, despite NGOs lobbying to have energy recovery demoted below recycling/composting on the waste management hierarchy, the government opted for an equivalent priority, as business had done (*ENDS Report* 1995, 251, p.15). The government also agreed heartily with business that, in principle, EC legislation was not needed and also that there should be greater emphasis placed on developing end-use markets for recyclate than was built into the German Ordinance.

There were two main differences between government and business. First, the government announced in November 1994 that a landfill levy would begin in late 1996 and would be collected by Customs and Excise under the auspices of the Treasury. This option had been dismissed by the PRG and other concerned sectors, e.g. the National Association of Waste Disposal Contractors (NAWDC), as raising revenue for the Treasury but not improving landfills or their environmental contribution (*ENDS Report* 1994, 238 p.3; in comparison, ACBE had supported such a levy from a very early stage). Secondly, the government continued to push responsibility onto industry, describing the development of recyclate markets as 'primarily the task of industry' (DOE/DTI 1994, p.11) and likewise the promotion of waste minimisation (*ENDS Report* 1994, 238, p.26).

Overall, the PRG's attempt at self-regulation had foundered and the government found itself picking up the pieces again and discussing its legislative options. When the government attempted to create and 'license' one business voice, it found that the issues were too wide to generate agreement. Secondly, neither the associations (through COPAC) nor the PRG were able to enforce compliance, nor did they wish to because packaging regulation found no ready business consensus. This is an interesting case where government, or more correctly one government Minister, looked to business to formulate and implement self-regulation because it seemed an easy way to deal with a 'new' issue by 'organising' it outside of central political responsibilities through the 'subpolitics' of business self-regulation. But this new issue drew on a diversity of sectors and companies and had little established organisation or coalition which might facilitate consensus (also Aguilar 1993).[8] So, instead of taking on voluntary self-regulation, business called for favourable regulation to defend UK business against European competition and to protect compliant companies and their investments from free-riders when legislation became inevitable (see Barrett 1991, p.11).

This example contrasts with the rationales for self-regulation presented earlier, which suggested that it serves to enhance legitimacy and pre-empt external regulation. In fact, the UK government's demands for packaging self-regulation were rejected in favour of more direct lobbying for regulatory protection, particularly of big companies and of the status quo. Perhaps we need to emphasise again that many of the business environmental activities given here and in previous chapters are more correctly reactive than proactive, especially where they react against a threat to the current system. And again, they seem to be dominated by large companies. Moreover, the packaging example underlines the privileged role of business groups in policy development. Although no self-regulation developed in this case, there was still a clear negotiation process between government and business that could have led to the implementation of public policy objectives through decentralised business organisations, if the problems of consensus had been resolved. Government thinking is more ideologically in line with business interests than with environmentalist NGOs, and the consultation culture reinforces this by making business access customary. Regardless of the eventual resolution of the packaging case, it is clear that it was to business that government turned for input, not to environmentalist NGOs, ensuring the continued if unpredictable business input to the national environmental agenda.

SUMMARY

This chapter has considered the variety of ways in which business influences the national agenda, especially through the mutually influential activities of self-regulation and lobbying. It is clear that self-regulation has important implications for legitimacy and influence on evolving legislation, especially where this is related to specialist business sectors with readily identifiable representatives. It is also clear that self-regulation is not guaranteed either legitimacy or influence, and often the very sectors which are most in need of these, being exposed to considerable public criticism or regulatory pressure, are going to find it most difficult to set up responses that are adequate in the eyes of the public and of governments.

NOTES

1. Of course, the environmental agenda is also influenced by those business activities discussed in previous chapters. For example, efforts in environmental reporting can either enhance the need for regulation or obviate it, thereby influencing political discussions and the prioritising of particular environmental issues. This chapter seeks to make explicit the impacts of these kinds of activities by showing how they are entangled with more proactive attempts to orient the wider environmental agenda in favour of business.
2. Readers should refer to the wider commentaries on this subject, e.g. Williamson (1989); Streeck and Schmitter (1985); Cawson (1985); Grant (1993).
3. The third ACBE will serve from 1996 to 1998 and be chaired by David Davies of Johnson Matthey. As yet, it has produced no documentation on which I might comment.
4. Of course, one of the main problems with investigating companies is that they are notoriously secretive and unwilling to allow researchers good access to all operations and information, especially where operations are sensitive and affect their public image, as environmental matters invariably do. It is possible that research into company environmental activities beyond the purely superficial level is much more difficult than that into operational management or personnel activities, because companies are far more wary of exposure to their critics.
5. COPAC's first plan (1992b) identified that packaging represented about 25% of domestic waste and therefore proposed a target of 'diverting from landfill' half of all used packaging by 2000 (i.e. about 12.5% of total domestic waste, COPAC 1992b, p.1). However, the target was qualified by arguing that even under a national scheme with 80% participation rates (very high), only 37% of used packaging could be recycled and, further, under 'American assumptions' of 60% participation (still very high) 21% could be recycled – which is only 5% of domestic waste, not 12.5% (COPAC 1992b, p.3).
6. This can be calculated from statistics on pp.19–20 of the plan.
7. This can be calculated from the statistics on p.19 and p.21 of the plan.
8. Additionally, environmental issues could be seen as peripheral for both business and (the Conservative) government – neither has a clear vision or consensus about their remediation. Packaging particularly was a 'hot potato' that neither wanted to pick up.

Business and the international environmental agenda

The previous chapter outlined the nature of business influence in the UK, where discretionary regulation and business self-regulation have encouraged the development of a number of overlapping business initiatives in the environmental field. Consequently, there is a closeness between government thinking and business attitude on environmental issues, which is partly founded on the peripheral nature of those issues to both parties at present and their incremental approach to solutions within a consensual framework.

In this chapter, I will turn my attention to the international scale, where there are similarities as well as differences to the UK case. Obviously, the range of interests encountered stretches not only across industrial sectors in one country but also across state boundaries, meaning that consensus must be built to embrace different political and regulatory traditions as well as operational backgrounds. In order to examine some of these issues, I shall consider the European scale briefly, which is similar in many ways to the UK, and then the more 'global' scale which, although dominated by western-born companies and viewpoints, looks to more strategic concerns and the longer-term development of the environmental agenda, rather than specific regulations.

EUROPEAN BUSINESS INFLUENCE ON ENVIRONMENTAL ISSUES

The European influence on national regulation has already been noted by examining the development of UK packaging policy in response to a draft EC Directive (Chapter 5 and more generally in Chapter 2). However, the influence of Europe, especially the European Commission, differs to that of national governments. We should always remember that EC legislation is rarely binding at the national level and must instead be formally introduced through national legislatures. It is therefore predominantly regulated by national agencies, notwithstanding the establishment of the European Environment Agency. This means that EC legislation tends to focus on setting targets and building in flexibility for national governments to decide on the means to achieve these targets. This leads to problems and conflicts where some states wish to push ahead of EC targets because they see them as too weak and others make only slow progress because they see them as too strict (Chapter 2). Business has been shown to lobby national governments in order to gain the support of national ministers in European debates, e.g. to argue for lower recycling targets at the EC level (Chapter 5). Hence, 'the growing importance of EC legislation may sometimes reinforce the dependency which exists at the national level between groups and "their" ministers' (Mazey and Richardson 1992, p.116). This also means that any study of business influence on the European environmental agenda must study national influences, as it is in effect national ministers who take policy decisions in Europe through the power of the Council of Ministers and the implementation of European decisions at national levels (McLaughlin *et al.* 1993; Haigh 1989).

In parallel to the UK case, most lobbying effort is directed specifically at the Commission, the civil service of the European Union, but the European Parliament is also an increasingly important target, especially with its enhanced powers under the Single European Act 1987 (Grant 1993). Despite criticisms in the UK, the EC's bureaucracy is remarkably small compared to that of the 12 national governments which it is supposed to monitor and deal with (Mazey and Richardson 1992, p.115). This can hamper its information-gathering and consultative processes on complex issues such as the environment.

In the 1980s, arrangements to centralise business lobbying of the European Commission were immature. This was reflected in the multiplicity of interest associations represented at the European level: nearly 300 for manufacturers and over 150 for commercial interests (Sargeant 1985, p.232). This made the Commission interested in developing and coordinating the roles of business associations and other NGOs in policy consultation and formulation (McLaughlin *et al.* 1993). This is because 'the expertise that can be provided by interest groups can help to move forward a policy process which can easily become overwhelmed by the volume of work that has to be done' (Grant 1993, p.168). In addition, the currently immature consultative processes are being consolidated with the explicit aim of developing more corporatist models in certain policy areas, which will favour business groups rather than other NGOs (Mazey and Richardson 1992, pp.116–26; see Chapter 5).

Research has produced widely differing views on the relative importance of lobbying bodies within the European consultation process at present. According to Mazey and Richardson (1992, p.124) at present key companies are 'the first port of call' for EC officials during consultation, but Grant (1993) argues that those officials increasingly favour European-wide associations rather than companies, because the former are more able to develop a European consensus. McLaughlin *et al.* (1993) suggest that there is no single answer to the question of the most important group. In their study of the car manufacturing sector, officials were predisposed to deal with European associations but turned increasingly to companies as problems arose within their negotiations. In order to achieve maximal influence, companies want to have both individual representation and input to the collective viewpoint of an appropriate association, because 'to pass over membership of a sectoral group is to throw away an important option' (McLaughlin *et al.* 1993, p.209).

An example of a European business association is the European chemical industry federation, CEFIC, 'one of the most effective industry associations operating at the EC level', which offers a single voice for EC officials to contact rather than 12 competing national associations (Grant 1993, pp.168–84). In the 1990s, the association developed a 'two-tier' system whereby both multinational corporations and a number of chemical associations from different sectors belonged directly to CEFIC. This enabled CEFIC to accommodate the different views and tactics of the two groups, to boost its funding (to around £5 million p.a.) and to enhance its consensus on policy, facilitated by the centralisation of the chemical sector (Mazey and Richardson 1992, p.118). CEFIC has produced specific 'Guidelines for the protection of the environment' as a model for companies in the chemical sector, based firstly on compliance but also on the requirement that companies 'take independent and responsible actions' beyond regulatory compliance (CEFIC leaflet, *CEFIC Guidelines for the Communication of Environmental Information to the Public*, n.d.).

However, other European-level business associations are more poorly resourced than CEFIC, certainly in comparison with their national equivalents. They may also have greater difficulty in building consensus across national borders (as we saw for sectoral borders in Chapter 5), especially in more diverse sectors such as food and drink (Mazey and Richardson 1992, p.118; Grant 1993, p.169). Where national associations and companies are reluctant to fund additional European groups, this affects resources further and can make their lobbying activity ineffectual as well as compromising their ability to present an authoritative business viewpoint to the Commission (McLaughlin *et al.* 1993). This is further undermined by the additional and separate lobbying by companies from a non-collective viewpoint. National associations may also take on direct European lobbying in their own interests, thereby duplicating or complementing the activities of related European associations. For example, the UK transsectoral business association, the CBI, funds a branch office in Brussels at a cost of £500 000 per year (Grant 1993, p. 170). The motor corporation Daimler has five so-called 'corporate embassies' in other countries and its Brussels one has double the staff of the Association of European Automobile Constructors, its European sectoral association (McLaughlin *et al.* 1993).

As well as sectoral business lobbies, there are cross-sectoral business groups at the European level, as at the national level. These include the Union of Industrial and Employers' Confederations of Europe (UNICE) and the European Round Table. The former is a peak association of national cross-sectoral associations, such as the CBI in the UK, and has had little input on environmental issues since it does not class them amongst its responsibilities (Grant 1993, p.175). Like the CBI, it suffers problems in building consensus across many different companies, leading to accusations of supporting only lowest common denominator opinions. The European Round Table has been more influential. Set up in 1983, with a Brussels secretariat, its invited membership of 45 chief executive officers (CEOs) presides over companies which collectively represented 60% of EU industrial production in 1991 (Doherty and Hoedeman 1994, p.135). Like ACBE (Chapter 5), these CEOs are 'business leaders' but they apparently enjoy better and closer contact with European Commissioners than does ACBE with national ministers. The Round Table was set up particularly to facilitate the establishment of the internal European market through reducing trade barriers. It therefore takes a free-market approach that is increasingly global in scope. Although critics have argued that the Round Table has become very influential and 'part of the EU apparatus' in some policy areas, e.g. transport, its voice specifically on the environment has been unobtrusive because it has only supported voluntary measures as a business response to environmental issues and not sought to influence policy in a concerted way (Doherty and Hoedeman 1994, p.139).

It is difficult to pick out clearly the impact of all these different business groups on the European environmental agenda. Overall, Mazey and Richardson (1992, p.122) argue that business groups have found DGXI (the environment Directorate-General of the EC) difficult to convince of their viewpoint, because it has been more in tune with environmental NGOs since its inception, more so than the Department of the Environment in the UK has been for example. Business has therefore turned to the other directorates for support. When proposals for a European carbon/energy tax emerged in 1991/2, although taken by surprise, business lobbyists developed links with DGXXI (which is responsible for indirect taxation) to emphasise the practical problems for business of the proposed tax (Ikwue and Skea 1994; Grant 1993, p.187). Those sectors were most active whose production costs would be most affected by such a tax—namely energy, iron and steel, cement and glass. They undertook 'a concerted attempt to kill the tax idea' by claiming it would damage business instead of its behaviour in a positive way (Ikwue and Skea 1994, p.5). The business argument was for a 'no regrets' approach, i.e. minimal expenditure on solutions in case the predicted (and uncertain) environmental problems did not materialise, a common approach also in UK policy. Business, via the CBI and ACBE in the UK, lobbied national ministers and the UK government came out in opposition to the tax. However, in the light of the UK government's predisposition against the Brussels administration as a perceived threat to national sovereignty, it would probably have taken this line in any case (Ikwue and Skea 1994, p.8). Other specific cases of European lobbying have revolved around the eco-label criteria (Chapter 4) where business lobbied for more favourable hurdles for their products through European sectoral associations, e.g. the European Tissue Symposium and the light bulb association (*ENDS Report* 1994, 234, p.23; 1994, 236, p.27; West 1995).

This brief excursion into European business lobbying illustrates the similarities to the UK case discussed in the previous chapter, namely the problems of cross-sectoral associations, the early but reactive influence on legislative drafts, the differential resourcing and foci of different associations and the importance of the formal channels to government or the Commission. Unfortunately, work on European business and environmental policy is as yet poorly documented and it is difficult to go beyond this brief outline. This is clearly an area where the increasing impact of the EC on the environmental agenda, and therefore on the expanding business lobby, needs to be matched by growing research attention and analysis.

BUSINESS AND THE GLOBAL ENVIRONMENT

However, if we turn our attention to the global scale, where lobbying is not necessarily focused on particular governments, we find some similar issues but also a more strategic and agenda-setting focus. It is here that we find the strongest potential for proactive influence at the broadest level and the international level is therefore the focus for the rest of this chapter.

I shall firstly outline two of the key players at this level, the ICC and the BCSD, which in many ways resemble the distinctions between the CBI and ACBE (noted in Chapter 5). Other sources of business influence include companies and the International Institute for Sustainable Development (IISD). The former (as noted in Chapter 5) leave a less conspicuous 'paper trail' of influence for us to monitor and it is probably therefore more helpful to consider the influence of multinational corporations through publications such as company policies (Chapters 3 and 4). The largest companies in particular have often set the blueprint for other companies and for the development of self-regulation on such issues. The IISD is a private non-profit-making corporation set up by the national government of Canada and the provincial government of Manitoba to promote sustainable development in decision making by governments, business and individuals. I shall consider how these groups affect the environmental agenda through looking specifically at the international debates about sustainable development and world trade, beginning by introducing the ICC and the BCSD.

The International Chamber of Commerce (ICC)

The ICC is main association for business at the international scale and it draws its membership from both manufacturing and service companies. The ICC is large and determinedly international, with over 7500 members in 110 countries, and provides information and seminar services as well as policies on specific issues drawn up by its own committees, known as Commissions. The ICC's general mission focuses on lobbying in order:

> to advance the interests of business throughout the world . . . [and] it is recognised officially as the spokesman [sic] for world business on a broad range of international policy matters. (ICC World Industry Council for the Environment Business Leaders' Forum on Sustainable Development factsheet, n.d., Appendix 2)

It therefore acts as a cross-sectoral business lobby at the international scale, aiming to represent general business interests, especially free trade, open markets and economic growth, and to have a policy influence.

Part of its role as a key voice is explicitly to speak on environmental issues. In 1989, the ICC UK Chairman claimed that 'the ICC is leading world industry' in the formulation of environmental policies (ICC UK Annual Report 1989, p.3[1]). In the late 1980s, international environmental interest was rising in such issues as sustainable development, prompted by the report from the Brundtland Commission (WCED 1987; see later) and the Montreal Protocol, resulting in greater media attention (Everett and Peplies 1992; Chapter 2). These pressures caused the ICC to consider how best to respond and to develop strategies for future involvement in the intensifying environmental debate. In 1990, it set up GEMI (the Global Environmental Management Initiative) under the US Council Foundation, the educational arm of its US body, the US Council for International Business. GEMI aimed to coordinate and share information about environmental performance standards, about ways of measuring and reporting across sectors to benefit multinational industry. It also aimed to have a proactive role in the business response to environmental issues.

> A center [such as GEMI] for corporate leadership and thinking on the subject of environmental management would substantially contribute to a proactive worldwide business ethic on the environment. (GEMI leaflet, n.d.)

The ICC launched its own code of environmental practice in its 'Business Charter for Sustainable Development' (hereafter 'the Charter'), produced as a free leaflet. This was developed with help from GEMI and officially adopted at the second World Industry Conference on Environmental Management (WICEM) in Rotterdam in April 1991. The aims of the Charter are threefold:

- to stimulate improved environmental performance;
- to guide environmental management; and, more implicitly,
- to legitimate business environmental activities to different publics and thereby pre-empt legislation.

These three activities are seen as mutually influential by the ICC as they will:

> demonstrate to governments and society that business is taking its environmental responsibilities seriously by helping to reduce the pressures on governments to over-legislate thereby strengthening the voice of business in public policy debates. (Willums and Golüke 1992, p.87)

The Charter contains 16 Principles to which companies pledge themselves when they sign up. These focus on management changes, such as increased environmental training for the workforce, research into the environmental impacts of industrial processes, and measuring and reporting environmental performance. The Principles are very positive but also very general, 'some might say vague' (Smart 1992, p.81), in that they set no targets for environmental performance, going only so far as to identify its dimensions. Eighteen months after the launch of the Charter, more than 1000 companies and business organisations had given it their formal support. Half of these were based in Europe and many were large multinationals: 50 were in the *Fortune* 500 list of the largest service companies as of July 1992 and a further 137 in the *Fortune* 500 list of manufacturing companies (ICC *Supporting Companies and Business Organizations* information, 1992). According to the ICC's figures, 36 out of the *Fortune* top 50 manufacturing companies are supporters of the Charter, again indicating that manufacturing companies are nearer the 'sharp end' of environmental issues because of their greater impact, higher (and therefore more exposed) profile and better resourcing (as we have seen throughout this book).

Despite this seemingly high level support for the ICC's Charter, its implementation remains voluntary. While it may enhance the environmental publicity and profile of both signatory organisations and the ICC, there is little retribution in cases where signatory companies do not implement the principles of the Charter. This has generated criticism of its effectiveness from many quarters. The President of the European Environmental Bureau (in Smart 1992, p.78) noted that the Charter must be properly implemented and its scope of activities expanded beyond internal management if companies and the Charter are to gain public credibility. The ICC had already realised this: the UK Chairman noted in 1991 that although the Charter was 'an outstanding milestone',

> expressions of support for the Charter from over 500 companies and organisations by the end of 1991 was not nearly enough. Evidence has to be provided to the governments participating at the Earth Summit in Rio in June that the admirable principles in the Charter are really working. Never has there been such a clear need for close collaboration between governments and industry world-wide as over environmental issues and the ICC is in a key position to ensure that happens in the years to come. (ICC UK Annual Report 1991, p.2)

As part of this process, the *ICC Charter* newsletter was set up in 1993 in order to report anecdotal experience of implementing the Charter, but there was still little standardised monitoring. The 1992 ICC book *From Ideas to Action* states that 'the ICC will not be monitoring compliance' with the Charter because it is inherently voluntary (Willums and Golüke 1992, p.88). However, in March 1992, in the run-up to the United Nations Conference on Environment and Development in Rio de Janeiro (UNCED or the 'Earth Summit'), the ICC and the United Nations Environment Programme (UNEP) convened an international meeting of ministers and industrialists, which agreed to set up a panel 'to periodically review progress in environmental management through the world-wide implementation of the Business Charter for Sustainable Development' (UNEP/ICC press release, 24 March 1992). Little has been seen of this panel since. In meetings after UNCED, the ICC made its position on voluntary adoption clear:

> The ICC will not monitor compliance with the Charter . . . The Charter is, rather a public commitment to a good faith process of improvement in environmental performance. Public interest will, in reality, be the monitoring mechanism. (quoted in UNEP IE/PAC report on the meeting, 9 September 1992, p.7)

Later, ICC UK suggested that 'we are now well into the process of encouraging our Charter supporters to implement the 16 Principles it contains' (ICC UK Annual Report 1993, p.2), again emphasising voluntary adoption rather than forced change. The Charter therefore remains only a code of conduct and an indication of the way in which business might change. In this sense, the ICC's input is to change the debate rather than force companies to implement specific improvements: its role is a strategic and representative one only and such a situation has been criticised. For example, Bebbington and Gray (1993, p.6) argue that regulations instead of voluntary measures are needed to make companies environmentally accountable and to audit for sustainability.

UNCED prompted the business community in general to take environmental issues even more seriously and to build a more strategic assessment of activities. The ICC, especially through its operative arm the International Environmental Bureau (IEB), was represented at Rio as part of the NGO forum, a presence that raised the hackles of other NGOs within the same forum who keenly felt the differences between business and themselves (*Earth Matters* 1992, 16, p.18). This representation formed part of a continuing campaign by the ICC to influence such international negotiations and seemed to be effective, at least on its own terms.

To contribute to its influence, the ICC produced a number of briefings for negotiators at Rio to make the business position plain (Table 6.1). As we might expect, these briefings advocated economic growth as the key to sustainable development, via open markets and the self-regulation of business, coupled with some harnessing of market forces in economic instruments of regulation. More succinctly, the ICC's recommendations have as a cornerstone 'support for free enterprise as a pre-condition for sustainable development' (ICC *Business Brief* 1992, 1, p.1). These briefings also see the complementarity and compatibility of economic and environmental imperatives as self-evident, whereas such an easy alliance has been debunked because of the philosophical divergences between business and environmentalist positions (Higham 1990b, p.17; see Chapter 1). Despite this, the ICC regarded its UNCED activities as successful.

> The fact that the outcome of the Earth Summit was generally favourable to business, more so than might have been expected, showed the benefit of much careful preparation and responsible input. The ICC provided a valuable cross-industry umbrella with credibility aided by the Business Charter and the global support behind

Table 6.1   Five principles in ICC lobbying at UNCED

---

1. Economic growth, open markets and environmental protection are complementary goals.
2. Sustainable development is an essential international goal that requires economic growth.
3. The private sector is the driving force for growth and environmental quality.
4. Sustainable development will make the best progress if it is set within a market economy.
5. Open trade is required for sustainable development because it encourages growth.

Source: adapted from Willums and Golüke 1992, p.18

it. (Peter Bright of Shell International, Environment Committee Chairman, ICC UK Annual Report 1992, p.8)

Andrew Lees, then Campaigns Director of Friends of the Earth (UK), highlighted the ICC's backing of sustainable development in public at UNCED, but claimed that 'one must decry the insidious way in which the ICC inflamed Northern politicians' fears about the recession and "reduced national competitiveness"' and how their emphasis on free trade was 'dressed up' for public consumption as sustainable development (*Earth Matters* 1992, 16, p.1). This again points to some success by the ICC in injecting the business point of view forcefully into the UNCED debate and in persuading governments to support its approach.

The ICC itself identified three specific fronts on which it was successful. First, calls for controls in order to protect the environment and promote development were weakened in *Agenda 21*, the main document produced by UNCED. There was, therefore, less impetus for the development of national regulations for environmental protection.

> We expect that these national laws and regulations will not be as stringent, bureaucratic and 'anti-business' as some feared before UNCED. (Willums and Golüke 1992, p.21)

Secondly, governments backed the business call for free trade within sustainable development, instead of NGO-supported arguments that free trade and environmental protection were not complementary goals. Thirdly, the sticky issue of the retention of intellectual property rights was resolved in favour of business (Willums and Golüke 1992, p.20).

Following UNCED, the ICC identified the need to consolidate its environmental lobbying channels. The result was WICE (World Industry Council for the Environment), approved in November 1992 and with an initial 1993 budget of US$1 million provided from membership fees. The rationalisation for this body was that:

> world business must have a strong, representative and proactive spokesman [sic] on sustainable development . . . [T]he future requires a strengthened capability to make a

major, ongoing policy impact through a form of operation which is CEO-led, issue-oriented, and agenda-setting. (ICC WICE Business Leaders' Forum on Sustainable Development, n.d., p.1)

Part of this proactive agenda setting is implicitly geared to the framing of the sustainable development debate (see later), explicitly through 'constructive lobbying' (p.2) and also implicitly through redirection of the debate towards company operational changes, rather than towards regulation.

> WICE's main task is to demonstrate and reinforce the business commitment to high environmental standards and make sure governments take full account of business recommendations as the agreements reached at the Earth Summit in 1992 are incorporated in national legislation. (ICC Annual Report 1993, p.2)

The Charter and WICE demonstrate that the ICC has made institutional changes in response to the international environmental debate, especially in the 1990s. But we should remember that, like most forms of self-regulation, its rationale is not necessarily to change company behaviour but to influence the views of the public and of regulators to be favourable to business. Again, this demonstrates that legislation and the pre-emption of legislation are prime reasons for (shallow) business environmental activity, at an international as well as a company scale.

The Business Council for Sustainable Development (BCSD)

The Business Council for Sustainable Development is a rather different body to the ICC and far younger. Maurice Strong, the Secretary General of UNCED, asked Stephan Schmidheiny (of the Swiss company UNOTEC) to act as a business adviser to the conference. Schmidheiny decided instead to set up the BCSD and asked about 50 'business leaders' to join him in order to put forward a coherent business view to the conference. The BCSD is therefore an invited group of individuals, not an association of associations with specific membership and services like the ICC, and it acknowledges this.

> The BCSD speaks not for global business but as a small group of business leaders, by definition representing a small miniority. We claim no legitimacy beyond our collective wisdom and that of the many people who have worked on this report. (Schmidheiny 1992, p.xxi)

Even so, the BCSD sees itself as an important group and, despite the long-established presence of the ICC and the generality of its activities, claims that:

> this is the first time that an important group of business leaders has looked at these issues from a global perspective and reached major agreements on the need for an integrated approach in confronting the challenges of economic development and the environment. (Schmidheiny 1992, p.xxi)

The first production of the BCSD was *Changing Course: A Global Business Perspective on Development and the Environment* (Schmidheiny 1992), following a series of working groups and workshops, published in time to influence UNCED in June 1992. It is instructive to look at this document as the coordinated viewpoint of the BCSD on a whole range of environmental and development-related topics. It intentionally provides only 'sketches' of how its ideas might be implemented, leaving the prescriptions deliberately vague (Schmidheiny 1992, p.xxii). The BCSD's general position is based on neo-classical economics, that free trade is the most important requirement of international policies and actions. This leads it to call for governments to provide 'open and accessible markets, more streamlined regulatory systems with clear and equitably enforced rules, sound and transparent financial and legal systems, and efficient administration' (p.xiii). Like the ICC, the BCSD sees markets in consumer sovereignty terms as ensuring individual freedom: 'freedom to participate in political decisions and freedom to participate in markets are inseparable over the long run' (p.9). Moreover, the BCSD sees the need for business to be more anticipatory and proactive in environmental activities.

> It is time for businesses to take the lead, because the control of change by business is less painful, more efficient, and cheaper for consumers, for governments, and for businesses themselves. (Schmidheiny 1992, p.83)

In *Changing Course*, the BCSD endorsed the general idea of economic incentives, where prices in open markets are 'made to reflect the costs of environmental as well as other resources' (p.14). This is a common business position: to prefer economic means to regulatory ones and argue that they are cheaper, more flexible and easier to administer. In practice, business often attacks specific proposals of this kind. For example, a carbon tax was only endorsed by the BCSD if it constituted a 'small initial carbon content charge' (not 'tax') (p.37) and was not unilaterally imposed. Furthermore, the BCSD qualified its approval of economic incentives by only supporting them where they are introduced 'slowly and predictably' and where they are revenue-neutral, also a common business stance. Hence, strategic ideas such as those promoted by the BCSD and others may appear positive, but the moment that they become a possible reality the business response is antagonistic.

> This raises the question whether business has truly 'changed course' or has simply developed a more sophisticated understanding of the way in which external pressures, particularly those of an environmental nature, influence its pursuit of traditional goals. (Ikwue and Skea 1994, p.8)

After UNCED, Schmidheiny chose to keep the BCSD going in a Phase II, 'to contribute to the post-UNCED agenda' through its own task forces and collaboration with other groups, e.g. the World Resources Institute and the Organisation for Economic Cooperation and Development (BCSD 1993, p.2). The independence of

Table 6.2   The fivefold mission of the BSCD

---

1. Provide business leadership as a catalyst for change towards sustainable development.
2. Work with and encourage policy makers to create the framework conditions required for business to make an effective contribution towards sustainable development.
3. Help to make business a respected partner in policy development and implementation.
4. Promote a clear understanding of sustainable development in the global business community and challenge business to self-examine its performance in this area.
5. Encourage business to develop goals and actions for sustainable development, within their current profit and loss criteria and also within the context of present and future international agreements, governmental policies and fiscal measures.

Source: adapted from BCSD 1993, p.3

the BCSD is still vaunted: 'it doesn't represent any sectoral interests, and it only speaks for itself' (BCSD 1993, p.8). It implicitly recognises its overlap with other organisations by arguing that its own 'leaders' want the debate to go faster and further than the lowest common denominator provided by more traditional cross-sectoral associations (i.e. the ICC). The five main aims of the post-UNCED BCSD are listed in Table 6.2 and trace similar notions to those of the ICC.

Thus, like the ICC, the BCSD's aims are both extravert and intravert because they recognise that business credibility amongst its regulators depends to a certain degree on the credibility of its internal operational and management changes. The public image and the private operations must therefore seem to be congruent to earn legitimacy. The BCSD has chosen to focus on 'eco–efficiency' and 'technology cooperation' as two key areas to encourage action in line with these aims. There is little evidence as yet of any input since UNCED by the BCSD and it is important to monitor its activities for signs of proactive influence on the environmental agenda, separate from its collaboration with other active bodies.

THE SUSTAINABLE DEVELOPMENT DEBATE

The refuge of the environmentally perplexed is sustainable development, namely wealth creation based on renewability and replenishment rather than exploitation. The trouble is that this is essentially a contradiction in terms for a modern capitalist culture. (O'Riordan 1989, p.93)

We can explore some of the issues noted above by analysing the effect of the ICC, the BCSD and the general business viewpoint that they express on a specific environmental example: the international sustainable development debate. First, I shall introduce the concept of sustainable development in a little more detail from an academic and environmentalist viewpoint, and then I shall assess the implications of the business viewpoint for the future direction of the agenda internationally.

The term 'sustainable development' has become the rallying cry from the reformist side of the environmental movement and particularly from advocates of 'green business'. Its basic premises are potentially radical but its adoption by a wide variety of environmental commentators has undermined its usefulness in the present environmental debate by diluting its impact. There are now many diverse interpretations in circulation, stemming from an 'absence of agreement about what exactly "sustainable development" means' (Redclift 1992b, p.34).

Although the notion of sustainable development has been around since the 1970s (Rees, 1991, p.294; Redclift 1987), it was first popularised internationally by the report of the World Commission on Environment and Development, known populatly as the Brundtland Commission after its chair (WCED 1987). The Brundtland definition has become the most quoted:

> Sustainable development seeks to meet the needs and aspirations of the present without compromising the ability to meet those of the future. (WCED 1987, p.40)

Variations on this theme are found frequently in policy and business statements.

The Brundtland Commission's report also set out seven 'critical objectives' for sustainable development policies. From a business perspective, three of these stand out: reviving growth to around 3% in developing countries and possibly 3 to 4% in developed countries if this is done sustainably; changing the quality of growth to make it less material and more equitable and energy efficient; reorienting technology towards innovation and environmental considerations (pp.50–52, 60). This analysis marked a departure from the more radical and pessimistic (for business) scenarios of global environmental management. It promised a positive view of environmental action, as opposed to the gloomy scenario of the 'limits to growth' debate which implied that growth and environmental quality were perpendicular goals (Pearce *et al.* 1989, pp.18–21; see Chapter 1).[2] In the Brundtland report, growth was not the problem but part of the solution: growth allowed the spreading of wealth and the undertaking of costly environmental management schemes. Hence the ICC's belief that sustainable development embodies 'the most important policy issue of our day directly affecting world business' (ICC WICE Business Leaders' Forum on Sustainable Development, n.d., p.4).

In stating that economic growth was not only inevitable but also necessary for good environmental management and north–south equity, the Brundtland Commission opened the door for business discussions on the topic. Previously, within a no- or low-growth framework, business had difficulties influencing the agenda in a positive light. This influence was invigorated when the Commission's report 'appeared to suggest that we do not need to sacrifice economic growth and real incomes in order to safeguard our environmental future' (Rees 1991, p.294). Happily for business, sustainable development could now be used to characterise pro-growth activities as simultaneously good for the environment, not unequivocally damaging. It therefore represented an opportunity for business rather than a threat, echoing the general repositioning of business positively in response to environmental issues more generally since the late 1980s.

> Sustainable development represents a major opportunity for companies. Those that rise to the challenge will position themselves strongly for the challenges of tomorrow's world. . . and, in doing so, dramatically improve their own long-term chances both of survival and success. (Deloitte Touche Tohmatsu International *et al.* 1993, p.50)

This positive view posited a compatibility between economic and environmental goals, so much so that, in the UK, it 'has now become fashionable for Government to embrace the concept of sustainable development and to endorse the notion that environmental and economic policies are inextricably intertwined' (Rees, 1991 pp.292–3; Chapter 1).

Sustainable development was also used by the Brundtland Commission to link economic growth and vitality to issues of poverty and inequality, especially across north–south divisions, and this theme was followed up at UNCED. In Principle 10 of the Rio Declaration produced following UNCED, we see a call for 'a supportive and open international economic system that would lead to economic growth and sustainable development in all countries'. Here are echoed the themes of inter-national equity, growth and environmental protection found in the Brundtland report. However, the continued emphasis on growth characterises environmental protection as requiring only the modification of economic systems, not fundamental changes in economies and institutions. Growth advocates take the view that 'poverty breeds environmental damage' (Buck, of the Institute of Directors, 1992, p.38). Sustainable development in this form is therefore a reformist notion (O'Riordan 1989, pp.93–4) and more acceptable to business than more radical no-growth arguments. In the 1990s, therefore, business can use sustainable develop-ment positively alongside its other pro-growth arguments rather than retreat to a defensive position on the topic.

Consequently, sustainable development has been employed by business in its environmental literature, but studies of this have been neglected in favour of analysing physical and technical changes to production, products and management (e.g. Elkington 1990; Elkington *et al.* 1991). I have already addressed some developments in relation to business use of sustainable development by assessing the ICC's response to the Brundtland report and UNCED 1992 and the estab-lishment of the BCSD. Of course, there are other bodies seeking to represent business at national and international levels: sectoral examples include the CIA in the UK and CEFIC (European chemical industry federation) at the European level; a transsectoral national example would be the CBI in the UK (see Grant 1993). All this literature reveals how business objectives have been implanted within the concept of sustainable development since WCED and UNCED. This is most clearly illustrated through the specific orientation of the ICC and the BCSD, because their strategic brief allows them more scope to address such wide agenda setting notions, and this orientation is worth examining.

Both the ICC and BCSD were very positive about the concept of sustainable development. The ICC's immediate response to WCED was to welcome the Brundtland report as a positive move and particularly to praise 'its emphasis on the importance of economic growth providing that the growth is sustainable' (ICC *Environment* Factsheet, February 1992). Both BCSD and IISD praise sustainable development as offering opportunities to business and making 'good business sense' (see Chapter 1). Suggestively, this business view considers sustainability to be a *secondary* aspect of economic growth, rather than justifying economic growth because it nurtures sustainable development as the Brundtland report seems to do.

> Economic growth provides the conditions in which protection of the environment can best be achieved, and environmental protection, in balance with other human goals, is necessary to achieve growth that is sustainable. (ICC in Smart 1992, p.78)

We see this even more clearly in the IISD report:

> It is important to understand and ensure that the sustainable development objectives that are established are ones that complement the enterprise's existing competitive strategies. In other words, sustainable development provides an additional dimension to business strategy. It provides senior management with an additional benchmark against which business strategies and performance should be assessed. (IISD and Deloitte Touche 1992, p.36)

The ICC's Business Charter for Sustainable Development uses the Brundtland definition given above without indicating that it is a direct quotation or embellishing on the definition. This implies that the ICC accepts the definition and therefore its overall arguments in favour of development in the widest sense as well as in the sense of growth. Similarly, both the BCSD in its main publication (Schmidheiny 1992, p.xi) and the IISD accept the definition in principle. However, the business groups employ contradictory ideas alongside it: the former rephrased WCED's arguments as being for 'rapid economic growth' without acknowledging that it may not necessarily be rapid or uniform (Schmidheiny 1992, p.3). The IISD (IISD and Deloitte Touche 1992, p.9) suggested that sustainability was necessary to ensure business survival, therefore reversing the priorities in the Brundtland argument for sustainable development. It went so far as to redefine the Brundtland definition specifically for business use:

> sustainable development means adopting business strategies and activities that meet the needs of the enterprise and its stakeholders today while protecting, sustaining and enhancing the human and natural resources that will be needed in the future. (IISD and Deloitte Touche 1992, p.11)

The report then employed a further contradiction when it bemoaned the fact that 'sustainable development is a concept that is not amenable to simple and universal definition'.

Divergences between the business and the Brundtland versions of sustainable development also appear when the ICC begins to operationalise the concept for use in its Charter. First, the ICC prioritises economic growth by citing 'growth' as what must be sustainable rather than the wider term 'development' with its social and political dimensions. In fact, development is often categorically referred to in the leaflet as 'economic development'. Secondly, the 16 principles of the ICC's Business Charter for Sustainable Development prioritise internal management changes, streamlining and progress measured against sectoral benchmarks. Although the Brundtland notion of sustainable development is referred to explicitly, issues of 'needs' and future generations are then bypassed in favour of environmental improvement and growth. The issues of equity, international equality and fair trade relevant to development in its fullest sense are not addressed, indeed are not mentioned in the Charter. This represents a specific use of the term sustainable development in order to further business efficiency and to prepare business for environmental criticism and developing regulation. It does not deal with the various forms of equity, particularly between north and south. To this extent, sustainable development is operationalised in terms of increased profits and reduced costs for the ICC's members, but also in terms of pre-emptive action by those members and the ICC as a body.

The implications of business influence on sustainable development

From the above, it is clear that the business groups made a big impact on the UNCED process, coming as it did after a period of preparation when they identified the issues and developed coherent responses from a business point of view. But the implications of their reorientation and specific operationalisation of sustainable development may be wider than UNCED. As noted in Chapter 5, business influence seeks not only to directly modify specific regulations but also indirectly to alter the agenda, public opinion and therefore the climate for future regulations. In this respect, the reorientation of sustainable development towards a business stance could have long-term implications for emerging international and national policies for sustainability. I shall briefly consider such effects in this section.

Business successes have been discussed generally by their critics as cynical and symbolic reorientations of the debate. Doherty and Hoedeman (1994, p.139) declare that the BCSD:

> cloaks all of its activities in 'greenwash'—unsubstantiated babble about the necessity of unregulated free trade in ensuring 'sustainable development', and glowing accolades of the environmental achievements of member companies.

This type of criticism can be deepened by analysing the business effects on three dimensions of the sustainable development debate: equity, technological change and markets.

Equity

We may assess the equity dimension of sustainable development on three levels. Most obviously, we have intergenerational equity, as cited in the Rio Declaration which sees 'eradicating poverty as an indispensable requirement for sustainable development'. This form of equity is encapsulated by 'futurity', accounting for the future instead of discounting it in economic analyses and also seeking to preserve environmental equality for future generations (Jacobs 1990, p.1; Pearce *et al.* 1989). This first level is the most publicly addressed by business groups such as the ICC and the BCSD. The issue of future generations, and therefore of maintaining long-term environmental quality, is used by business both as a rationale for technical environmental management and as an emotional tool for encouraging 'green consumer' expansion (e.g. use of children in environmentally oriented car advertising, Holder 1991; see Chapter 4).

Secondly, we have intragenerational equity across groups and classes, related to structural assessments of the differential impacts of environmental degradation for different groups (Pearce *et al.* 1989, pp.28–37; Morrison and Dunlap 1986). This second level is more difficult to assess, and this is not the place to undertake a full analysis. However, there is little explicit acknowledgement of class differences in business deployment of sustainable development as yet and nothing about how business activities might address them. This is despite analyses showing that business has employed market-skimming practices, where a high premium is set on environmentally marketed products (Simms 1992), therefore reinforcing class differences in the market.

Thirdly, we have international equity, particularly emphasising the development aspect of sustainable development through its effects on north–south inequalities and dependencies (e.g. Redclift 1987, 1992a,b). The business use of sustainable development has largely ignored this third form of equity, despite the fact that large transnational members of the ICC contribute to north–south differences (Gorz 1980, pp.114–30; Athanasiou 1996). International equity is implicit in the ICC's arguments for free international trade, but the issue is not one of evening out privilege and deprivation but of removing barriers to competition. This can reinforce inequalities because it fails to recognise that 'the wealthier countries of the world "import sustainability" from the poorer countries' (Pearce *et al.* 1989, p.47). The IISD (IISD and Deloitte Touche 1992, p.33) does acknowledge the importance of the 'alleviation of poverty and distributional equity', but there is little suggestion of how this might be dealt with at present, and development and the environment prove secondary to strategic business concerns in the majority of its report. Moreover, the ICC and the BCSD seem disposed to suppress these kinds of arguments in the sustainable development debate so that they can protect the large companies from which they draw their support.

> The BCSD became famous during the UNCED process for ensuring that language supporting the regulation of transnational corporations was cut from final declarations. (Doherty and Hoedeman 1994, p.139)

Overall, therefore, the equity issues identified in the Rio Declaration are neglected by business in favour of growth issues in the north. Therefore, this orientation of sustainable development does not address key sources of global environmental problems. In consequence, this undermines the success of sustainable development because 'sustainability and equity are two sides of the same coin; unless the latter is confronted many countries will simply not have the resources to meet the requirements of environmental sustainability' (Rees 1991, p.303). The use of sustainable development by business implies a misuse or at least a partial use of its meaning.

Techno-fix

The second implication that business use of sustainable development has is for the level of change. An overwhelmingly technical orientation of change is displayed by business, dependent on product and process amelioration and investment in research and development, e.g. in the ICC's Charter. Technological innovation is constructed as the means to improve environmental performance and reach environmental standards. Major lifestyle changes and the fundamental restructuring of economies, institutions and the debt regime are not on this agenda. The commitment to growth or free trade is predicated on the continued operation of the established economic system (Schmidheiny 1992). Thus, 'sustainability will seemingly be achieved through technological change which will allow gradual and relatively minor changes to occur in the developmental growth path with minimal economic pain' (Rees 1991, p.303).

So, the business emphasis on growth and technical environmental management belies key issues of sustainable development that would necessitate more fundamental examination and restructuring of national and global economies. Environmental management systems of the kind promoted by the ICC's Charter do not address sustainable development. Instead, they attempt to enhance business credibility for the benefit of business. Moreover, such stances now have a strong grip on government and the ideology of business environmental change, so that this orientation will be difficult to displace.

> The key concept of sustainable development requires a new approach to business . . . Indeed it has been argued that current approaches are suboptimal and inappropriate but that they are still likely to be adopted widely because they will become part of a dominant ideology. A responsible and pro-active approach to the environment requires new and radical approaches to doing business. This will include the need for increasing not decreasing legislation. (Welford 1993, p.32)

Welford's approach would require a very different approach to sustainable development by business nationally and internationally. Superficially, the IISD (1992, p.12) does at least recognise a difference between environmental management as a short-term change and sustainable development as a long-term strategy for global business. It appreciates the need for the social aspects of development to be addressed in the south (IISD 1992, p.33), but it still places this within the traditional business viewpoint as dependent on free trade above all else. Again, the status quo is supported and reinterpreted for the sustainable development debate.

Market-fix

The third implication is for business regulation. The ICC and the BCSD clearly push for self-regulation and, if this is not enough for governments, they back the principle of market-based and financial incentives for environmental change, e.g. pollution taxes and permits (see Pearce *et al.* 1989; Pearce 1992; Helm 1991). This fits into ideologies of western capitalism and of neo-conservatism through non-intervention in markets and the centrality of individual choices under business autonomy. Market-based instruments work within established operations by altering economic priorities across a range of activities to favour environmentally sound options, not by restricting business options to environmentally sound but costly ones. The emphasis is on the flexible response of business to an economic incentive and penalty system, because 'tax solutions are the natural outcome' of neo-classical economic reasoning (Gray 1990, p.143). Again, this orientation precludes a wider debate about restructuring the global economy, because it permits only modification to be considered. Like technical changes, the changes are shallow and reformist, looking not to equity or morality but to the accurate economic weighting of decisions. However, this general support of economic intervention is not necessarily followed in practice when specific economic measures are proposed by policy makers. As noted above, the BCSD supported the general notion of economic incentives as preferable to regulation but had only a lukewarm response to a specific carbon tax. The ICC claimed that a carbon tax would reduce private investment and distort international trade. Consequently, it preferred a 'no-regrets' policy to respond to fears of climate change in order to avoid 'drastic short-term change that might be economically damaging' (ICC Annual Report 1993, p.7).

This ambivalence reflects the reality that business groups support economic incentives as preferable to regulatory options, but not as preferable to voluntary action with no external intervention at all. The accepted free-market ideology leads business groups to reject international controls devised in the name of sustainable development. The ICC and WICE argue that 'open international trade is the most effective way to support global economic growth which *in turn* is essential for sustainable development', again rendering sustainable development as a secondary consideration (ICC Annual Report 1993, p.2, emphasis added). In this way, sustainable development may be 'a constraint on the market economy' (Jacobs 1990, p.11) but not a fundamental reorientation of it. Again, the emphasis on growth and economic evaluation obscures issues of equity. So, business use of sustainable development has been criticised because of its reorientation of critical environmental and developmental issues.

> While in no way disputing that market mechanisms could produce a more *efficient* means of achieving emission targets, the resulting distributions of environmental capacity and emission control costs need be neither *equitable* nor politically acceptable. (Rees 1991, p.299, emphasis in original).

Effects on the debate

Overall, the sustainable development debate is clearly being influenced by business input. The very plasticity of the term has made that input more influential in the 1990s (and see Athanasiou 1996). Redclift points out that sustainable development brings together two opposing intellectual traditions—limits to growth and human potential—producing 'more than a pious hope, but rather less than a rigorous analytical schema' for appropriate policy decisions (Redclift 1987, p.198). This is easily influenced by business to emphasise growth with added but shallow environmental improvements. In this way, the contradictions within the concept of sustainable development (growth versus equity versus environmental protection) are obscured (Redclift 1987, p.2) and the environmental debate is reframed by business around growth and free trade, deemphasising equity and restructuring. This powerful reorientation must therefore preclude the discussion of policy options, essentially those which would curb business autonomy, and by this strategy business seeks to maintain its autonomy and market control.

There is much debate about how the concept of sustainable development, and its sibling term 'sustainability', might be operationalised and therefore monitored. If business can make an effective input at this stage, it has the potential to bring the concept into line with its own practices in a proactive way. As noted in Chapter 5, this would allow business to develop a blueprint which can then be offered to regulators and international policy makers as a successful way to operationalise the concept, perhaps using the codes of conduct and weak self-regulatory practices of the ICC's Charter. The IISD makes this evident when it argues that, because there is no clear definition, the 'norms' of sustainable development are currently being established through business *practices* rather than policy discussions: 'companies are ahead of government in establishing sustainable development performance criteria' (IISD 1992, p.41). The international bodies are clearly urging this route on business because of its potential to influence the wider environmental debate proactively.

> The business community should not wait for such standards to be developed before individual enterprises experiment with setting sustainable development objectives and measuring and reporting on the sustainability of their enterprises. (IISD 1992, p. 41)

The use of sustainable development by business is therefore different to its use by development commentators and environmental NGOs. Because of the power of the business lobby internationally, this business-specific view can be forcefully articulated at high levels of decision making and influence the direction of the environmental agenda and consequent policy, especially where that orientation demotes environmental action to only shallow and reformist dimensions.

BUSINESS, INTERNATIONAL TRADE AND THE ENVIRONMENT

As well as concerns about sustainable development, prompted by international environmental and development concerns, environmental issues have been linked to the most central concern of international business: trade. A focus for this discussion has been the General Agreement on Tariffs and Trade (GATT), set up in 1948 to remove trade barriers in international markets. The Uruguay Round of GATT negotiations began in 1986 and was due to end in 1990, but it dragged on until 1993 because of disagreements over socially and environmentally related barriers to trade. This greatly concerned the business groups and multinational corporations especially, who are of course particularly active in the international business lobby. Both business and the environmental NGOs have taken the GATT, and the Uruguay Round especially, as a focus for their trade arguments and exerted much pressure on this important global issue.

> Never before has our world business organisation campaigned with such insistence and for so long in support of an international agreement . . . Had the Round failed, all nations would have been the losers. (ICC Annual Report 1993, p.2)

The GATT is based on free-market ideology and increasingly looks to the expanding markets in the east and south as new areas to be opened to global competition under its rules (or rather its rejection of rules). Many of the arguments used for and against the GATT are those underlying the deep divergences between business and environmentalist groups (see Chapter 1). The global business view is that 'free trade has a role to play in the progress toward sustainable development' (Schmidheiny 1992, p.69). This view believes that the GATT should prioritise free trade above all else and not allow any other agreements to take precedence, except certain multilateral ones which do not distort trade. The GATT was never designed to address environmental concerns (Beacham 1992) and business will accordingly fight to ensure that it cannot now 'be transformed into an agreement that puts more emphasis on environment than on trade' (Schmidheiny 1992, p.74).

For critics, of course, this is precisely the fault of the GATT. They focus on 'the bottom-rung position which both environmental protection and quality of life issues take in the turmoil of establishing political and economic security' in the global economy so that 'environmental sustainability becomes one more barrier to competitiveness' (Poff 1994, p.443). Critics of the GATT, especially NGOs, would prioritise many things before trade.

> The needs of people and the environment, in rich and poor countries, come before trade concerns, and the rules of the General Agreement need to be revised to enable individual nations to take progressive, protective action. (Friends of the Earth 1992, p.6)

The business viewpoint sees threats like these to the GATT as threats to trade, which imply a nightmare vision of 'a drift away from the multilateral trading system towards a world of managed trade, thinly-disguised discrimination and outright protectionism' (ICC Annual Report 1993, p.2). It caricatures the critics of GATT as taking 'a rather simplistic view that economic growth necessarily hurts the environment, and trade must therefore be bad because it spurs economic growth' (Schmidheiny 1992, p.72), without providing evidence for the reverse, i.e. the equally 'simplistic' business view that trade does *not* hurt the environment. For example, Schmidheiny (1992, p.73) also suggests that the 'international trading system is rule-based, not power-based', an analysis that many commentators on development and the environment would totally reject.

Critics of the GATT and the free-market ideology in general have long pointed out that the current trade system is not multilaterally equal. In some senses, world trade is already 'managed' due to the influence of the multinational corporations on trade patterns (Williams 1993, p.93). The GATT merely further protects trade and the positions of the dominant countries in international trade patterns (Beacham 1992, p.681). Moreover, protectionism is well established, particularly in agricultural policy in Europe. There is no evidence that national environmental policies will have any detrimental effects on trade patterns, because 'the share of environmental costs in output value is too small to be an important component in firms' decision making' (Williams 1993, p.85). This underlines the fact that this conflict is a clash of deeply held philosophies and structural perspectives, not a clash of empirically based arguments.

> The GATT may have achieved its goals too well, because we are now left with a system that is entrenched, inflexible, and unaccommodating of environmental concerns. (Beacham 1992, p.667)

A corollary of the anti-protectionist argument forwarded by business groups is that open markets ensure economic and political freedom. Both the BCSD and the ICC make the claim that 'market access is surely one of the most effective ways of helping these nations [in the south] to achieve economic well-being and ward off the danger of political extremism' (ICC Annual Report 1993, p.2). However, critics seek to expose the immanent links between economic power and social and political power. The most glaring flaw in the business argument is the hypothesis that economic processes drive all other benefits, when decades of studies have exposed the inequality of the debt regime and shown that claims for 'value-free' economics have no basis. Friends of the Earth (UK) has argued that the GATT is 'morally bankrupt' because it ignores power inequalities in economic patterns of trade. Demands from environmental NGOs that environmental considerations should be prioritised above trade have caused their opponents in turn to claim that this counts as 'eco-imperialism' in similarly ignoring the power inequalities, this time not in trade but in the national capacity to deal with environmental problems. In other words, environmental policies would force poor countries to invest in environmental technologies at the expense of development and poverty relief. Additionally, NGOs have argued that the entire GATT process is secretive and unaccountable, giving little information about how its decisions are reached and allowing no participation by other groups (Friends of the Earth 1992).

Clearly, business has the greatest horror of unilateral measures and protectionism because they threaten to curtail or reorient trade. Business instead argues that environmental damage is not the result of a lack of regulation but of regulation which interferes with market operations. Consequently, the removal of such protectionism will be good for trade and, in a natural progression, good for the world (see Williams 1993, p.89). However, critics have argued that, in the witch hunt for economic protectionism, the GATT simultaneously denies national sovereignty and any democratic rights which have been built into national constitutions to protect their own citizens and environments by forcing countries with high environmental standards to relax them in the name of trade.

> Transnational corporations have a political and undemocratic message that citizens in all nations have to be competitive and that that is to be accomplished by dismantling national institutions, social programs and environmental protections. The fact that competitiveness without the protection of our natural resources, our infrastructure and social programs amounts to mass suicide is rarely considered. (Poff 1994, p.444)

A strict interpretation of the GATT may threaten even agreed international policies if these employ trade restrictions to force changes to behaviour. For example, the Montreal Protocol calls on signatory countries not only to stop their own CFC production but also not to import CFCs from other countries that do not meet the Protocol requirements, except those with 'grace periods' before implementation, e.g. China and India. Hence, the Montreal Protocols ask countries to do precisely what the GATT does not want them to do: to use trade restrictions to protect the environment (Beacham 1992, p.672). Countries which implement environmental (and other) legislations contrary to the GATT's provisions may therefore face international tribunals and resultant penalties.

In the end, and perhaps inevitably given the powers ranged in its support, the Uruguay Round was concluded without environmental issues being incorporated into its rules. Although this was a huge relief to business interests, it is clear that this omission will lead to further discussions internationally about the connections between trade and the environment. Business groups have committed themselves to fighting any further so-called environmental protectionism:

the ICC will certainly do its utmost to prevent environmental, social or any other issues, however justified in themselves, from becoming a pretext for protectionism. (ICC Annual Report 1993, p.2)

The fundamental disagreements over the GATT, therefore, are due to divergent ideologies, a business one which supports trade above all else and a critical one which denies that absolute priority in favour of political, social and environmental considerations (see Chapter 1). The conflict will not be resolved without a restructuring of one of these ideologies, which is highly unlikely at this time. Hence,

nothing short of fundamental reform of the GATT will allow effective measures to save the global environment from further destruction . . . [because the GATT] . . . is currently the greatest obstacle to the formation and enforcement of international agreements and domestic policies aimed at protecting the global environment. (Beacham 1992, pp.679–81)

SUMMARY

This chapter has looked at the international business influence through concentrating on the activities and ideologies of the ICC and the BCSD. It is clear that their foci are more strategic than those noted in the national case, i.e. sustainable development and the GATT, and these form a context for the national activities and legislation with which we have been concerned so far. In consequence, the influence at the international level is more general: consensus is built despite sectoral differences because of the wide scope of the issues unlike those relating to the specifics of national regulations.

But it is probably the international and more general level than can generate the greater impact for business on the environmental agenda in the long term. The real power seems to lie in aligning the widest concepts to business thinking in the stages which precede national policy formulation, and therefore helping to formulate the way in which that policy will address environmental issues. If business is becoming more positive and proactive about international and global environmental issues, then we should look even more carefully at how those issues are worked out in agreements and how far the solutions subsequently offered match business requirements more closely than NGO requirements. It is in the framing of the environmental agenda and not its intimate details that business may yet have the greatest and least environmentally friendly influence.

NOTE

1. This is despite the fact that the environment had not been mentioned in previous annual reports.
2. For the limits to growth debate, see Meadows et al. (1972); Hardin (1977); Maddox (1972); O'Riordan (1981).

7

Environmental issues and business: some implications

Welcome to the new world, where everything is changing, and changing not at all.
(Athanasiou 1996, p.4)

In the last decade or so, business has sought to respond to environmental issues and their associated social, political and regulatory pressures. Its response embraced operational change, corporate policies, advertising campaigns, lobbying efforts and self-regulation, as we have seen. Whilst this variety makes it difficult to generalise across sectors, company sizes and corporate environmental attitudes, it does seem clear from the preceding chapters that business seeks to respond to these pressures in order to ensure its legitimacy and financial viability. Sometimes the legitimacy side of the equation is emphasised, as in the chemical industry's Responsible Care programme (Chapter 5), sometimes financial viability is emphasised, as in the call to good environmental 'housekeeping' (Chapter 1). In reality, the two intertwine and form a powerful rationale for business to be seen to be responding to environmental issues. The large companies in particular are positioning themselves as the 'pioneers' in environmental response, because their higher public and governmental profiles expose them more to environmental pressures and, often, they have larger investments in environmental technologies and markets which they seek to protect.

In the previous chapters, I have tried to analyse the range of business responses and to demonstrate that in most cases they support the status quo, thus representing 'reformist' or shallow change. The business viewpoint on this contrasts strongly with the environmentalist critique, which seeks a far more radical overhaul of economic and political relations at national and international levels. The environmentalist critique of business environmental activity is therefore negative, in terms of what business does not do or what it does badly, whereas within business the discussion has sought out the positive, the business potential to offer environmental solutions, particularly since the dissemination of the Brundtland report

(Chapter 6) and the development of green consumerism (Chapter 2) in the late 1980s. Hence, environmental issues have been cast as 'opportunities' not 'threats' for business and commentators have fallen over themselves to urge business to develop coherent, legitimate and well-founded responses.

This raises the issue of who initiates business environmental activity. Although business professes proactive environmental change, it is clear that much change remains dependent on some form of outside pressure, especially regulation. Business environmental activity is therefore mainly reactive, with smaller companies especially going little further than compliance with economic and legislative forces, but larger companies coupling this with additional anticipation and forward planning. So, whilst prescriptive frameworks (Chapter 3) call for more interactive, proactive and creative environmental activity, business is still concentrating on the current and imminent pressures of its situation. This suggests that the most effective forcing mechanisms will continue to be external pressures, particularly legislation which has impacts on all corporate sizes and cultures.

There is clearly a need for further academic research into many of these issues. For example, as codes of practice and regulatory schemes develop in the 1990s, their impacts must be traced and their relative importance in exerting pressures on business generally and within different sectors must be evaluated. At the time of writing, this process has only just begun. Without, therefore, a great deal of literature on which to draw, I shall use this final chapter to suggest some general implications of the business response to environmental issues. The key question is, if environmental considerations become part of the business mainstream, the 'new orthodoxy' as some suggest (e.g. Burke 1990), what does this mean for the future of the environmental agenda? While necessarily tentative, these implications point to areas for future research attention as well as paint scenarios which may or may not be realised, depending on how we as society and as individuals choose to respond to business environmental activity.

IMPLICATIONS FOR MARKETS

Business environmental activity has influenced markets and economic consider-ations through product development, marketing, pricing, information provision and availability.

This has two dimensions. Firstly we have its influence on levels of environmental damage due to production and consumption and the spread of 'green consumerism' in the late 1980s, particularly in Europe. This has been claimed by some as a success for NGOs in gaining wider support, but alternatively by others as a success for business in developing new markets. It is clear that business has responded to social trends towards environmental concern through its product ranges, packaging and advertising. But, unlike the NGOs which campaign for product boycotts or reduced overall consumption under the 'green consumerism' banner, business sees opportunities in a large green market because more consumers would be involved but the technical and lifestyle changes that they adopt would not make established markets redundant. Growth is implied, not contraction.

This contrasts two possibilities for the future of the 'green market': a minority of people with very different and less environmentally damaging lifestyles or a majority of people with slightly changed behaviour. The former will not require a strong green market developed by multinational companies, it will display lower overall consumption and prefer local shops, healthfood stores and home-grown produce; but the latter would both require and support a large green market with modified rather than radically redesigned products and processes. Clearly, business would favour the latter. Further, we cannot empirically prove which option might be the more environmentally sound. 'Greening' the market in the second manner will not necessarily have any effect on overall environmental impact because more products will be sold, even if each one has slightly less impact. In this view, 'green consumerism' and 'green business' endorse the current capitalist ideology because they reduce intervention, play on individual responsibility and ownership and offer little challenge to the economic orthodoxy (Harte *et al.* 1990; Pepper 1989/90; Irvine 1989). Again, it is a reformist change to the market. The obvious corollary to this is that business, in its own interests, would wish to encourage mainstreaming of green concerns where this means that it gains markets and sales. Over the next few years, it will be fruitful to study how business is seeking to develop such a trend and how this is demonstrated in the nature of the green market. Even more importantly, we should monitor (if we can) whether 'green consumerism' and 'green business' are increasing per capita consumption of products, resources and energy, and consider whether this alone exposes the paradox between their surface arguments and their underlying ideologies.

Secondly, as well as overall consumption, the 'greening' of the market implies a focus on economic action so that it eclipses other types. For example, might most of the new converts to environmental concern be content to buy 'green' products and not seek to support environmental organisations, get involved in environmental campaigns, reduce their overall consumption or consider voting for environmental reform? Might new converts only support the more reformist aspects of the green agenda and not more radical lifestyle and societal change? I suspect the answer to both questions is yes. In consequence, if economic action becomes the predominant channel for protest, people's ability to use other channels might atrophy (Hirschman 1970). It is true that other forms of protest are being kept alive in the UK, at least at present, by radical and collective grass roots activity, especially against road construction, and that the environmental movement in all countries has always embraced more radical critiques of economic and political systems which have consistently attacked all kinds of businesses for their paltry environmental credentials. But there remains the danger of polarisation, that the mainstream will accept economic channels of environmental expression but not political (or 'lifestyle') ones. This threatens the legitimacy of the environmental movement and its support base in the future, although the fact that it currently maintains much higher public trust than business should protect it somewhat.

IMPLICATIONS FOR REGULATION

We might expect business environmental activity to have similar implications for regulation, in the sense that the discussion of environmental solutions will focus on economic arguments and actions. As I have illustrated, public and political pressures on business are closely linked because often legislation takes its cue from public concern or, in turn, influences and directs that concern. Hence, business influence on regulation, and relatedly on legislation and political support, would seem to have similar implications to its influence on markets. Where environmental policy is developed, we might expect business to seek to emphasise economic and also technological solutions rather than social or structural ones. For example, business would support the adoption of catalytic converters in private cars rather than the restructuring of and consistent investment in public transport networks. The industrial claim to specialist knowledge is particularly important in this regard in swaying regulators about what is technically and financially possible and in gaining influence over the details of regulation and the mechanisms to achieve legislative goals.

Perhaps more important than the move towards economic and technological regulatory solutions is the attempt by business to avoid regulation entirely. Galbraith (1972) claims that autonomy is a prime goal for business, perhaps more important than profit. Certainly, Chapters 2, 5 and 6 detailed business attempts to pre-empt regulation by offering self-regulatory blueprints (Responsible Care), by internalising the issue (COPAC) and by claiming achievements in advance of regulatory implementation (the ICC's Charter; BAMA and CFC control). All these mechanisms would guarantee greater business autonomy than external regulation, and therefore more flexibility to respond to pressures. Clearly, if this kind of business influence is to succeed, it depends on considerable legitimation as well as a favourable political and economic climate. At present, it seems that environmental regulation is one area which is unlikely to contract and all the signs are that more is inevitable, particularly from the European Commission.

However, we have also seen cases where business wants regulation. The packaging case in Chapter 5 demonstrates that self-regulation is not always satisfactory for business interests, particularly where it might affect competition or where consensus is difficult to build. In such cases, large companies especially would rather protect their own 'pioneering' investments than take part in collective action to bring competitors up to their own level.

In either case, it is still likely that business has a much greater influence on the final form of regulation than other NGOs. It draws on greater and more stable resources, can profess specialist knowledge and an important role in the economy, and it is customarily consulted with more alacrity than other lobbies.

IMPLICATIONS FOR THE ENVIRONMENTAL AGENDA

What both these sets of influences imply is that the overall influence of business on the environmental agenda could be substantial. Environmental solutions are likely to be reformist and only economically and technologically justified. More radical lifestyle changes or the restructuring of sociopolitical relations become more difficult as they challenge the orthodoxy which is shared by business and government. This may well be a 'new green orthodoxy', but it is unlikely to be radical or to move too far from the mainstream of economic and business thinking. Partly this is a trade-off between broad but shallow 'greening' or deep but narrow change. It seems a truism that for environmentalism to increase its spread and participation, and to be adopted by a variety of businesses and consumers, it must become diluted.

The most dangerous implication is that this new orthodoxy could gain dominance over the environmental agenda and therefore constrain its development. As Bebbington and Gray (1993) have noted, where one mode of thinking dominates so strongly, others are silenced before debate can begin. It is a moot point whether such a dominance has yet developed and whether it is unmovable, but it is clear that NGOs recognise this threat and continually work against the closing down of environmental policy avenues by business arguments. One of the reasons that I have often assessed the proactivity in business action is that it can indicate how far business is seeking to reorient the agenda. In most cases, proaction is bound up with reaction to impending and inevitable pressures, but Chapter 6 does offer some wider suggestions of the way in which proaction might work to change the international environmental agenda in favour of business. The redefinition and operationalisation of sustainable development in economic and technological terms obscure the currently pressing political and economic inequalities between nations and social groups.

Such reorientation is very general and does not guarantee an echo in corporate operations and planning. This underlines the necessity of looking not only to what business is actually doing and its physical environmental impact, but to what business is saying and appearing to be and its impact on the environmental agenda in political, social and ideological terms. It is the latter that might prove far more significant in the long term.

Bibliography

ACBE (1991) *First Progress Report* (DTI/DOE, London)

ACBE (1992) *Second Progress Report* (DTI/DOE, London)

ACBE (1993a) *Report of the Financial Sector Working Group* (DTI/DOE, London)

ACBE (1993b) *Third Progress Report* (DTI/DOE, London)

ACBE (1994) *Fourth Progress Report* (DTI, London)

ACBE (1996) *Sixth Progress Report* (DTI/DOE, London)

ACBE/DOE/DTI (1993) *A Guide to Environmental Best Practice for Company Transport* (DOE, London)

ACCA/CBI (1994) *Introducing Environmental Reporting: Guidelines for Business* (CBI, London)

Adams, Richard (1992) 'Green reporting and the consumer movement' 106–18 in David Owen (ed.) *Green Reporting: Accountancy and the Challenge of the Nineties* (Chapman and Hall, London)

Adams, Richard, Jane Carruthers and Sean Hamil (1990) *Changing Corporate Values: A Guide to Social and Environmental Policy and Practice in Britain's Top Companies* (Kogan Page/New Consumer, London)

Aguilar, Susan (1993) 'Corporatist and statist designs in environmental policy: the contrasting roles of Germany and Spain in the European Community scenario' *Environmental Politics* 2, 2, 223–47

Ashford, Nicholas A (1993) 'Understanding technological responses of industrial firms to environmental problems: implications for government policy' 277–307 in Kurt Fischer and Johan Schot (ed.) *Environmental Strategies for Industry: International Perspectives on Research Needs and Policy Implications* (Island Press, Washington DC)

Athanasiou, Tom (1996) 'The Age of Greenwashing' *Capitalism Nature Socialism* 7, 1, 1–36

B&Q (1995) *How Green Is My Front Door?* (B&Q, Eastleigh, UK)

Baggott, Rob (1989) 'Changing face of self-regulation' *Public Administration* 67, 435–54

Ball, Simon and Stuart Bell (1991) *Environmental Law* (Blackstone Press, London)

Ball, Simon and Stuart Bell (1994) *Environmental Law* 2nd edition (Blackstone Press, London)

Barrett, Scott (1991) 'Environmental regulation for competitive advantage' *Business Strategy Review* Spring, 1–15

BCSD (1993) *Business Council for Sustainable Development* information leaflet (BCSD, Geneva)

Beacham, K Gwen (1992) 'International trade and the environment: implications of the General Agreement on Tariffs and Trade for the future of environmental protection efforts' *Colorado Journal of International Environmental Law and Policy* 3, 655–82

Bebbington, Jan and Rob Gray (1993) 'Corporate accountability and the physical environment: social responsibility and accounting beyond profit' *Business Strategy and the Environment* 2, 2, 1–11

Beck, Ulrich (1992) *Risk Society* (Sage, London)

Bendall, Jem and Francis Sullivan (1996) 'Sleeping with the enemy? Business–environmentalist partnerships for sustainable development. The lessons of the WWF 1995 Group' presented at the Business–Environmentalist Partnerships: A Sustainable Model Conference, Cambridge, UK.

Benedick, Robert Elliot (1991) *Ozone Diplomacy: New Directions in Safeguarding the Planet* (Harvard University Press, Cambridge MA)

Biddle, David (1993) 'Recycling for profit: the new green business frontier' *Harvard Business Review* November–December, 145–56

Birkin, Frank and Helle Bank Jørgensen (1994) 'Tales in two countries: an insight into corporate reporting in Denmark and the UK' *Business Strategy and the Environment* 3, 3, 10–15

Blaza, Andrew (1992) 'Environmental reporting: a view from the CBI' 31–4 in David Owen (ed.) *Green Reporting: Accountancy and the Challenge of the Nineties* (Chapman and Hall, London)

Blowers, Andrew (1984) *Something in the Air: Corporate Power and the Environment* (Harper and Row, London)

Boddewyn, J J (1985) 'Advertising self-regulation: organization structures in Belgium, Canada, France and the United Kingdom' 30–43 in Wolfgang Streeck and Philippe C Schmitter (eds) *Private Interest Government: Beyond Market and State* (Sage, London)

British Organic Farmers, Organic Growers Association and Safeway plc (1991) *Organic Fact File: A Guide to the Production and Marketing of Organic Produce in the United Kingdom* (BOF/OGA/Safeway plc, London)

Bronze, Lewis, Nick Heathcote and Peter Brown (1990) *The Blue Peter Green Book* (BBC Books, London)

Buck, Janice (1992) 'Green awareness: an opportunity for business' 35–9 in David Owen (ed.) *Green Reporting: Accountancy and the Challenge of the Nineties* (Chapman and Hall, London)

Burke, Tom (1990) 'Did we go green in the '80s?' *Town and Country Planning* 59, 1, 11–12

Burke, Tom and Julia Hill (1990) *Ethics, Environment and the Company: A Guide to Effective Action* (Institute of Business Ethics, London)

Burnside, Amanda (1990) 'Keen on green' *Marketing* May 17, 35–6

Buttel, Frederick H (1986) 'Discussion: economic stagnation, scarcity, and changing commitments to distributional policies in environment-resource issues' 221–38 in Allan Schnaiberg, Nicholas Watts and Klaus Zimmerman (eds) *Distributional Conflicts in Environment-Resource Policy* (Gower, Aldershot)

Buttel, Frederick H and Oscar W Larson III (1980) 'Whither environmentalism? The future political path of the environmental movement' *Natural Resources Journal* 20, 2, 323–44

Cannon, Tom (1994) *Corporate Responsibility: A Textbook on Business Ethics, Governance, Environment: Roles and Responsibilities* (Pitman, London)

Carey, Anthony (1992) 'A questioning approach to the environment' 87–97 in David Owen (ed.) *Green Reporting: Accountancy and the Challenge of the Nineties* (Chapman and Hall, London)

Carruthers, Jane (1993) 'Flushed with success: Shell's management of an oil spill' 50–63 in Malcolm McIntosh (ed.) *Good Business? Case Studies in Corporate Social Responsibility* (Centre for Social Management, School for Advanced Urban Studies/New Consumer, Bristol)

Cawson, Alan (ed.) (1985) *Organized Interests and the State* (Sage, London)

CBI (1986) *Clean Up—It's Good Business* (CBI, London)

CBI (1990) *Waking Up to a Better Environment* (CBI, London)

CBI (1994) *Britain's Business Voice* (CBI, London)

CBI/ICC (1990) *Environmental Auditing* Conference transcript (Rooster, Royston, Herts)

CIA (1992a) *Responsible Care* (CIA, London)

CIA (1992b) *The Chemical Industry and the Environment: An Agenda for Progress* (CIA, London)

CIA (1993) *The New Agenda* (CIA, London)

Clarke, D B and M G Bradford (1989) 'The use of space by advertising agencies within the United Kingdom' *Geografiska Annaler* **71B**, 3, 139–51

Coleman, William (1990) 'State traditions and comprehensive business associations: a comparative structural analysis' *Political Studies* **38**, 231–52

Coleman, William and Wyn Grant (1988) 'The organizational cohesion and political access of business: a study of comprehensive associations' *European Journal of Political Research* **16**, 467–87

Coopers and Lybrand Deloitte (1990) *Going Organic: The Future for Organic Food and Drink Products in the UK* (Coopers and Lybrand Deloitte, Birmingham)

COPAC (1992a) COPAC Business Plan to Address UK Integrated Solid Waste Management (COPAC)

COPAC (1992b) COPAC Action Plan to Address UK Integrated Solid Waste Management (COPAC)

Corbett, Charles J and Luk N Van Wassenhove (1993) 'The green fee: internalizing and operationalizing environmental issues' *California Management Review* Fall, 116–35

Cosgrove, D (1994) 'Contested global visions: *One-world, Whole-Earth* and the Apollo space photographs' *Annals of the Association of American Geographers* **84**, 2, 270–94

Cotgrove, Stephen (1982) *Catastrophe or Cornucopia: The Environment, Politics and the Future* (John Wiley, New York)

Crepaz, Markus M L and Arend Lijphart (1995) 'Linking and integrating corporatism and consensus democracy: theory, concepts and evidence' *British Journal of Political Science* **25**, 2, 281–8

Daniels, Stephen (1993) *Fields of Vision: Landscape Imagery and National Identity in England and the United States* (Polity Press, Cambridge)

Datschefski, Edwin (1992) *Environmental Sense is Commercial Sense* (Business and the Environment Programme, The Environment Council, London)

deHaven-Smith, Lance (1988) 'Environmental belief systems: public opinion on land use regulation in Florida' *Environment and Behaviour* **20**, 2, 176–99

Deloitte Touche Tohmatsu International, International Institute for Sustainable Development and SustainAbility (1993) *Coming Clean: Corporate Environmental Reporting: Opening Up for Sustainable Development* (Deloitte Touche Tohmatsu International, London)

Devall, Bill (1990) *Simple in Means, Rich in Ends: Practising Deep Ecology* (Green Print, London)

Dewhurst, Philip (1990) 'Environmental crisis: CFCs and the ozone layer: how ICI handled a major public issue' 64–78 in Danny Moss (ed.) *Public Relations in Practice: A Casebook* (Routledge, London)

Dobson, Andrew (1990) *Green Political Thought* (Unwin Hyman, London)

DOE (1990) *This Common Inheritance: Britain's Environmental Strategy* (HMSO, London)

DOE (1992) *This Common Inheritance: The Second Year Report* (HMSO, London)

DOE/DTI (1994) *Recycling: The Government's Response to the Second Report from the House of Commons Select Committee on the Environment* (HMSO, London)

Doherty, Ann and Oliver Hoedeman (1994) 'Misshaping Europe: the European Round Table of Industrialists' *The Ecologist* **24**, 4, 135–41

Downs, Anthony (1972) 'Up and down with ecology: the "issue-attention cycle"' *Public Interest* **28**, 38–50

Doyle, Jack (1992) 'Hold the applause: a case study of corporate environmentalism' *The Ecologist* **22**, 3, 84–90

Drotning, Philip T (1972) 'Why nobody takes corporate responsibility seriously' *Business and Society Review* Autumn, 68–72

DTI (1994) *Guidelines for Non-advertising Green Claims* (DTI, London)

Dudek, Daniel J, Alice M LeBlanc and Kenneth Sewall (1990) 'Cutting the cost of environmental policy: lessons from business response to CFC regulation' *Ambio* **19**, 6–7, 324–8

Eckersley, Robyn (1992) *Environmentalism and Political Theory: Toward an Ecocentric Approach* (UCL Press, London)

Eckersley, Robyn (1993) 'Free market environmentalism: friend or foe?' *Environmental Politics* **2**, 1, 1–19

Eden, Sally E (1992) *Individual Motives and Commercial Retailing in Green Consumerism* Unpublished Ph.D. thesis (Leeds University, Leeds)

Eden, Sally E (1993a) 'Individual environmental responsibility and its role in public environmentalism' *Environment and Planning A* **25**, 1743–58

Eden, Sally E (1993b) 'Constructing environmental responsibility: perceptions from retail business' *Geoforum* **24**, 4, 411–21

Eden, Sally E (1994) 'Using sustainable development: the business case' *Global Environmental Change* **4**, 2, 160–7

Eden, Sally E (1995) 'Business, trust and environmental information: perceptions from consumers and retailers' *Business Strategy and the Environment* **3**, 4, 1–8

EEO/ACBE (1993) *Practical Energy Saving Guide for Smaller Businesses: Save Money and Help the Environment* (EEO/DOE, London)

Einsiedel, Edna and Eileen Coughlan (1993) 'The Canadian press and the environment: reconstructing a social reality' 134–49 in Anders Hansen (ed.) *The Mass Media and Environmental Issues* (Leicester University Press, Leicester)

EIRIS (Ethical Investment Research Service) (1993) *Attitudes to Ethical Investment: A Study of Institutional Investors in GEC, Hanson and ICI* (EIRIS, London)

Elkington, John (1989) 'Going greener about the gills' *Marketing Week* **12**, 38, November 24, 56–60

Elkington, John (1990) *The Environmental Audit: A Green Filter for Company Policies, Plants, Processes and Products* (World Wide Fund for Nature, London)

Elkington, John (1994) 'Towards the sustainable corporation: win–win–win business strategies for sustainable development' *California Management Review* Winter, 90–100

Elkington, John and Tom Burke (1987) *The Green Capitalists* (Victor Gollancz, London)

Elkington, John with Anne Dimmock (1991) *The Corporate Environmentalists. Selling Sustainable Development: But Can They Deliver?* (SustainAbility Ltd, London)

Elkington, John and Julia Hailes (1988) *The Green Consumer Guide* (Victor Gollancz, London)

Elkington, John and Peter Knight with Julia Hailes (1991) *The Green Business Guide* (Victor Gollancz, London)

Epstein, Edwin M (1981) 'Societal, managerial, or legal perspectives on corporate social responsibility: product and process' 81–103 in S Prakash Sethi and Carl L Swanson (eds) *Private Enterprise and Public Purpose: An Understanding of the Role of Business in a Changing Social System* (John Wiley, New York)

Everett, Michael D and Robert Peplies (1992) 'The political economy of environmental movements: US experience and global movements' *Environmental Values* **1**, 297–310

Ewen, Stewart (1976) *Captains of Consciousness: Advertising and the Social Roots of the Consumer Culture* (McGraw-Hill, New York)

Fessenden-Raden, June, Janet M Fitchen and Jenifer S Heath (1987) 'Providing risk information in communities: factors influencing what is heard and accepted' *Science, Technology and Human Values* **12**, 3&4, 94–101

Fischer, Frank (1993) 'The greening of risk assessment: towards a participatory approach' 98–115 in Denis Smith (ed.) *Business and the Environment: Implications of the New Environmentalism* (Paul Chapman, London)

Fletcher, Frank (1989) 'Selling to the greens? It's not so simple' *What's New In Marketing* August, 20–2

Flynn, Andrew and Philip Lowe (1992) 'The greening of the Tories: the Conservative Party and the environment' 9–36 in Wolfgang Rüdig (ed.) *Green Politics Two* (Edinburgh University Press, Edinburgh)

Forrester, Susan (1990) *Business and Environmental Groups: A Natural Partnership* (Directory of Social Change, London)

Friedman, Milton (1988) 'The social responsibility of business is to increase its profits' 217–23 in Thomas Donaldson and Patricia H Werhane (eds) *Ethical Issues in Business: A Philosophical Approach* (Prentice Hall, New York)

Friends of the Earth (1992) *Fools' Gold: The General Agreement on Tariffs and Trade and the Threat of Unsustainable Development* (Friends of the Earth, London)

Galbraith, John Kenneth (1972) *The New Industrial State* (André Deutsch, London)

Gallisot, Laurent (1994) 'The cultural significance of advertising: a general framework for the cultural analysis of the advertising industry in Europe' *International Sociology* **9**, 1, 13–28

Gardner, Carl and Julie Sheppard (1989) *Consuming Passion: The Rise of Retail Culture* (Unwin Hyman, London)

Giddens, Anthony (1990) *The Consequences of Modernity* (Polity Press, Cambridge)

Gilbert, Mike (1994) 'BS 7750 and the eco-management and audit regulation' *Eco-Management and Auditing* **1**, 2, 6–10

Goll, Irene and Gerald Zeitz (1991) 'Conceptualizing and measuring corporate ideology' *Organization Studies* **12**, 2, 191–207

Goodin, Robert E (1992) *Green Political Theory* (Polity Press, Cambridge)

Gorz, André (1980) *Ecology as Politics* (South End Press, Boston MA)

Gorz, André (1989) *Critique of Economic Reason* (Verso, London)

Grant, Wyn (1983) 'Representing capital: the role of the CBI' 69–84 in Roger King (ed.) *Capital and Politics* (Routledge Direct Editions, London)

Grant, Wyn (1984) 'Large firms and public policy in Britain' *Journal of Public Policy* **4**, 1, 1–17

Grant, Wyn (1987) 'Introduction' 1–17 in Wyn Grant (ed.) *Business Interests, Organizational Developments and Private Interest Government: An International Comparative Study of the Food Processing Industry* (de Gruyter, Berlin)

Grant, Wyn (1992) 'Models of interest intermediation and policy formation applied to an internationally comparative study of the dairy industry' *European Journal of Political Research* **21**, 53–68

Grant, Wyn (1993) *Business and Politics in Britain* 2nd edition (Macmillan, Basingstoke)

Grant, Wyn and David Marsh (1977) *The CBI* (Hodder and Stoughton, London)

Grant, Wyn, Alberto Martinelli and William Paterson (1989) 'Large firms as political actors: a comparative analysis of the chemical industry in Britain, Italy and West Germany' *West European Politics* **12**, 2, 72–90

Gray, Rob (1990) *The Greening of Accountancy: The Profession after Pearce* (Chartered Association of Certified Accountants, London)

Gray, Rob, Dave Owen and Keith Maunders (1987) *Corporate Social Reporting: Accounting and Accountability* (Prentice Hall, New York)

Gray, Rob, Dave Owen and Roger Adams (1995) 'Standards, stakeholders and sustainability: the ACCA Environmental Reporting Awards 1994' *Certified Accountant* March, 1–5

Green Party Manifesto (1987) *Green Party General Election Manifesto* (Heretic Books, London)

Greenpeace (n.d.) *The Greenfreeze Story* (Greenpeace, London)

Haigh, Nigel (1989) *EEC Environmental Policy and Britain* (Longman, Harlow)

Hansen, Anders (1993) 'Greenpeace and press coverage of environmental issues' 150–78 in Anders Hansen (ed.) *The Mass Media and Environmental Issues* (Leicester University Press, Leicester)

Hardin, Garrett (1977) 'The tragedy of the commons' 16–30 in Garrett Hardin (ed.) *Managing the Commons* (Freeman, San Francisco)

Harte, George and David Owen (1991) *Environmental Disclosure in the Annual Reports of British Companies: A Research Note* (University of Leeds School of Business and Economic Studies Discussion Paper AF91/04, Leeds)

Harte, George, Linda Lewis and David Owen (1990) *Ethical Investment and the Corporate Reporting Function* (University of Leeds School of Business and Economic Studies Discussion Paper AF90/07, Leeds)

Harvey, David (1974) 'Population, resources, and the ideology of science' *Economic Geography* **50**, 3, 256–77

Heberlein, T A (1974) 'The three fixes: technological, cognitive and structural' in D R Field, J C Barron and B L Long (eds) *Water and Community Development: Social and Economic Perspectives* (Ann Arbor Science, Ann Arbor MI)

Heelas, Paul and Bronislaw Szerszynski (1991) 'Byuing the right stuff' *Town and Country Planning* **60**, 7, 210–11

Helm, Dieter (1991) *Economic Policy Towards the Environment* (Blackwell, Oxford)

Hemming, Christine (1992) 'The proposed EC eco-audit scheme: a pilot exercise' *European Environment* **2**, 3, 9–11

Higham, Nick (1990a) *Marketing Week* **13**, 15, June 22, 17

Higham, Nick (1990b) *Marketing Week* **13**, 19, July 20, 17

Hill, Claude (1986) *The Future for Organically Grown Produce* (Food From Britain, London)

Hill, Julie (1992) *Towards Good Environmental Practice: A Book of Case Studies* (Institute of Business Ethics, London)

Hilton, Anthony (1989) 'The importance of being Green' *Marketing* August 31, 15

Hirschman, Albert O (1970) *Exit, Voice and Loyalty: Responses to Decline in Firms, Organisations and States* (Harvard University Press, Cambridge MA)

Hoffman, Andrew J (1994) 'The environmental transformation of American industry: an institutional account of environmental strategies in the chemical and petroleum industries' presented at the Greening of Industry conference, Copenhagen, Denmark

Holder, Jane (1991) 'Regulating green advertising in the motor car industry' *Journal of Law and Society* **18**, 3, 323–46

Hopper, Joseph R and Joyce McCarl Neilsen (1991) 'Recycling as altruistic behaviour: normative and behavioural strategies to expand participation in a community recycling program' *Environment and Behaviour* **23**, 2, 195–220

Hoskins, W G (1970) *The Making of the English Landscape* (Penguin, Harmondsworth)

Hutchinson, Colin (1992) 'Corporate strategy and the environment' *Long Range Planning* **25**, 4, 9–21

Ikwue, Tony and Jim Skea (1994) 'Business and the genesis of the European community carbon tax proposal' *Business Strategy and the Environment* **3**, 2, 1–10

IISD (International Institute for Sustainable Development) and Deloitte Touche (1992) *Business Strategies for Sustainable Development: Leadership and Accountability for the '90s* (IISD, Winnipeg)

IOD (1992) *Stewards of the Earth: Environmental Policy for a Market Economy* (IOD, London)

IOD (1993) *IOD Members' Opinion Survey: Environment* (IOD, London)

IOD (1995) *IOD Business Opinion Survey: The Environment* (IOD, London)

Irvine, Sandy (1989a) *Beyond Green Consumerism* (Friends of the Earth, London)

Irvine, Sandy (1989b) 'Consuming fashions? The limits of green consumerism' *The Ecologist* **19**, 13, 88–93

Irvine, Sandy and Alec Ponton (1988) *A Green Manifesto* (Macdonald Optima, London)

ISBA (Incorporated Society of British Advertisers Ltd) (1992) *Environmental Claims in Advertising: A Single Guide to All the Applicable Advertising Codes* (ISBA, London)

Jacobs, Michael (1990) *Sustainable Development: Greening the Economy* (Fabian Society, London)

Johnston, R J (1989a) 'An environmentalist upsurge in Great Britain?' *Environment and Planning A* **21**, 851–2

Johnston, R J (1989b) *Environmental Problems: Nature, Economy and State* (Belhaven Press, London)

Jordan, Grant (1991a) 'The professional persuaders' 13–46 in Grant Jordan (ed.) *The Commercial Lobbyists* (Aberdeen University Press, Aberdeen)

Jordan, Grant (1991b) 'Effective lobbying: the hidden hand' 173–89 in Grant Jordan (ed.) *The Commercial Lobbyists* (Aberdeen University Press, Aberdeen)

Jordan, Grant (ed.) (1991c) *The Commercial Lobbyists* (Aberdeen University Press, Aberdeen)

Karrh, B W (1990) 'Du Pont and corporate environmentalism' 69–76 in W Michael Hoffman, Robert Frederick and Edward S Petry (eds) *The Corporation, Ethics and the Environment* (Quorum Books, New York)

Keman, Hans and Paul Pennings (1995) 'Managing political and societal conflict in democracies: do consensus and corporatism matter?' *British Journal of Political Science* **25**, 2, 271–81

Kemp, Penny and Derek Wall (1990) *A Green Manifesto for the 1990s* (Penguin, London)

Kemp, René (1993) 'An economic analysis of cleaner technology: theory and evidence' 79–113 in Kurt Fischer and Johan Schot (eds) *Environmental Strategies for Industry: International Perspectives on Research Needs and Policy Implications* (Island Press, Washington DC)

Klatte, Ernst (1991) 'Environmental and economic integration in the EEC' 37–67 in Owen Lomas (ed.) *Frontiers of Environmental Law* (Chancery, London)

Knight, Alan P (1996) 'A Report on B&Q's 1995 Timber Target' presented at the Business–Environmentalist Partnerships: A Sustainable Model Conference, Cambridge, UK

Lampkin, Nicholas (1990) *Organic Farming* (Farming Press, Ipswich)

Lipworth, Sir Sydney (1994) *Packaging Levy Report on Point of Funding* (PRG, London)

Lowe, Philip and Jane Goyder (1983) *Environmental Groups in Politics* (Allen & Unwin, London)

MacEwen, Ann and Malcolm MacEwen (1987) *Greenprints for the Countryside? The Story of National Parks* (Allen and Unwin, London)

Macve, Richard and Anthony Carey (1992) *Business, Accountancy and the Environment: A Policy and Research Agenda* (Institute of Chartered Accountants in England and Wales, London)

Maddox, John (1972) *The Doomsday Syndrome* (Macmillan, London)

Manzo, L C and N D Weinstein (1987) 'Behavioural commitment to environment protection: a study of the active and inactive members of the Sierra Club' *Environment and Behaviour* **19**, 6, 673–94

Mazey, Sonia and Jeremy Richardson (1992) 'Environmental groups and the EC: challenges and opportunities' *Environmental Politics* **1**, 4, 109–28

Mazur, Allan and Jinling Lee (1993) 'Sounding the global alarm: environmental issues on the US national news' *Social Studies of Science* **23**, 681–720

McCormick, John (1989) *The Global Environmental Movement* (Belhaven, London)

McCormick, John (1991) *British Politics and the Environment* (Earthscan, London)

McCormick, John (1995) *The Global Environmental Movement* 2nd edition (Wiley, Chichester)

McCulloch, Alistair (1990) 'Mirror, mirror on the wall: who's the greenest of us all?' 210–18 in Wolfgang Rüdig (ed.) *Green Politics One* (Edinburgh University Press, Edinburgh)

McEvoy, James III (1972) 'The American concern with the environment' 214–36 in William R Burch Jr, Neil H Cheek Jr and Lee Taylor (eds) *Social Behavior, Natural Resources and the Environment* (Harper and Row, New York)

McLaughlin, Andrew, Grant Jordan and William A Maloney (1993) 'Corporate lobbying in the European Community' *Journal of Common Market Studies* **31**, 2, 191–212

Meadows, D H, D L Meadows, J Randers and W W Behrens III (1972) *The Limits to Growth* (Earth Island, London)

Meadows, Dennis L and Donella H Meadows (1973) *Toward Global Equilibrium: Collected Papers* (Wright-Allen, Cambridge, MA)

Miller, Alan (1992) 'Green investment' 242–61 in David Owen (ed.) *Green Reporting: Accountancy and the Challenge of the Nineties* (Chapman and Hall, London)

Miller, Charles (1991) 'Lobbying: the development of the consultation culture' 47–64 in Grant Jordan (ed.) *The Commercial Lobbyists* (Aberdeen University Press, Aberdeen)

Miller, Van and John Quinn (1993) 'How green is the transnational corporation?' *Business Strategy and the Environment* **2**, 1, 13–25

Morrison, Denton E, Kenneth E Hornback and W Keith Warner (1972) 'The environmental movement: some preliminary observations and predictions' 259–79 in William R Burch Jr, Neil H Cheek Jr and Lee Taylor (eds) *Social Behavior, Natural Resources and the Environment* (Harper and Row, New York)

Morrison, Denton E and Riley E Dunlap (1986) 'Environmentalism and elitism: a conceptual and empirical analysis' *Environmental Management* **10**, 5, 581–9

Nash, Tom (1990) 'Green about the environment?' *Director* **43**, 7, 40–44

Offe, Claus (1981) 'The attribution of public status to interest groups: observations on the West German case' 123–58 in Suzanne Berger (ed.) *Organizing Interests in Western Europe: Pluralism, Corporatism and the Transformation of Politics* (Cambridge University Press, Cambridge)

O'Riordan, Timothy (1976) *Environmentalism* (Pion, London)

O'Riordan, Timothy (1981) *Environmentalism* 2nd edition (Pion, London)

O'Riordan, Timothy (1989) 'The challenge for environmentalism' 77–102 in Richard Peet and Nigel Thrift (eds) *New Models in Geography* (Unwin Hyman, London)

O'Riordan, Timothy and Steve Rayner (1991) 'Risk management for global environmental change' *Global Environmental Change* **1**, 4, 91–108

Ottman, Jacqueline A (1992) 'Industry's response to green consumerism' *Journal of Business Strategy* **13**, 4, 3–7

Owen, David (1992) 'The emerging green agenda: a role for accounting?' 55–74 in Denis Smith (ed.) *Business and the Environment: Implications of the New Environmentalism* (Paul Chapman, London)

Pearce, David (1992) 'Green economics' *Environmental Values* **1**, 1, 3–13

Pearce, David, Anil Markandya and Edward B Barber (1989) *Blueprint for a Green Economy* (Earthscan, London)

Pearce, Fred (1992) 'Corporate shades of green' *New Scientist* October 3, 21–2

Peattie, Ken and Moira Ratnayaka (1992) 'Responding to the green movement' *Industrial Marketing Management* **21**, 103–10

Pepper, David (1989/90) 'Green consumerism—Thatcherite environmentalism' *New Ground* Winter, 18–20

PERI (Public Environmental Reporting Initiative) (1994) *PERI Guidelines* (PERI/BP, London)

Piesse, Jennifer (1992) 'Environmental spending and share price performance: the petroleum industry' *Business Strategy and the Environment* 1, 1, 45–54

Poff, Deborah C (1994) 'Reconciling the irreconcilable: the global economy and the environment' *Journal of Business Ethics* 13, 439–45

Porritt, Jonathon (1984) *Seeing Green: The Politics of Ecology Explained* (Blackwell, Oxford)

Porritt, Jonathon and David Winner (1988) *The Coming of the Greens* (Fontana, London)

PRG (1994a) *Real Value from Packaging Waste: A Way Forward* (PRG, London)

PRG (1994b) *Real Value from Packaging Waste* (PRG, London)

PSC (Packaging Standards Council) (1993) *Report of the Packaging Standards Council* (PSC, Datchworth, Herts)

Purvis, Martin (1994) 'Yesterday in Parliament: British politicians and debate over stratospheric ozone depletion, 1970–92' *Environment and Planning C: Government and Policy* 12, 361–79

Purvis, Martin, Frances Drake, David Clarke, Deborah Philips and Amatsia Kashti (1995) *Fragmenting Uncertainties: British Business and Responses to Stratospheric Ozone Depletion* (School of Geography University of Leeds Working Paper 95/12, Leeds)

Redclift, Michael (1987) *Sustainable Development: Exploring the Contradictions* (Methuen, London)

Redclift, Michael (1992a) 'The meaning of sustainable development' *Geoforum* 23, 3, 395–403

Redclift, Michael (1992b) 'Sustainable development and global environmental change: implications of a changing agenda' *Global Environmental Change* 2, 1, 32–42

Rees, Judith (1991) 'Equity and environmental policy' *Geography* 76, 4, 292–303

Roberts, Peter (1995) *Environmentally Sustainable Business: A Local and Regional Perspective* (Paul Chapman, London)

Robins, Nick (1990) *Managing the Environment: The Greening of European Business* (Business International, London)

Rolston, Holmes III (1989) *Philosophy Gone Wild: Environmental Ethics* (Prometheus Books, Buffalo)

Roome, Nigel (1992) 'Developing environmental management strategies' *Business Strategy and the Environment* 1, 1, 11–24

Roome, Nigel and Sarah Clarke (1994) 'Exploring the sustainable enterprise: a journey through theory and practice' presented at the Greening of Industry Conference, Copenhagen, Denmark

Rose, Chris (1990) 'Perception and deception: the collapse of the green consumer' presented at Wildlife Link White Paper 1990 Conference, York, UK

Ross, Andrew (1991) 'Is global culture warming up?' *Social Text* 28, 3–30

Rüdig, Wolfgang (1992) 'Editorial' 1–8 in Wolfgang Rüdig (ed.) *Green Politics Two* (Edinburgh University Press, Edinburgh)

Rüdig, Wolfgang and Philip Lowe (1986) 'The withered "greening" of British politics: a study of the Ecology Party' *Political Studies* 34, 262–84

Samdah, D M and R Robertson (1989) 'Social determinants of environmental concern: specification and test of the model' *Environment and Behaviour* 21, 1, 57–81

Sandbach, Francis (1978) 'Ecology and the "Limits to Growth" debate' *Antipode* 10, 2, 22–32

Sargeant, Jane (1985) 'Corporatism and the European Community' 229–53 in Wyn Grant (ed.) *The Political Economy of Corporatism* (St Martin's Press, New York)

Schjaer-Jacobsen, Hans and Hans Bundgaard (1994) 'On the transformation of sustainability requirements into industrial competitive advantage' presented at the Greening of Industry Conference, Copenhagen, Denmark

Schmidheiny, Stephan (with the Business Council for Sustainable Development) (1992) *Changing Course* (MIT Press, Cambridge MA)

Schmitter, Philippe C (1985) 'Neo-corporatism and the state' 32–62 in Wyn Grant (ed.) *The Political Economy of Corporatism* (St Martin's Press, New York)

Schnaiberg, Allan (1980) *The Environment: From Surplus to Scarcity* (Oxford University Press, New York)

Sethi, S Prakash (1981) 'A conceptual framework for environmental analysis of social issues and evaluation of business response patterns' 69–80 in S Prakash Sethi and Carl L Swanson (eds) *Private Enterprise and Public Purpose: An Understanding of the Role of Business in a Changing Social System* (John Wiley, New York)

Sethi, S Prakash (1990) 'Corporations and the environment: greening or preening?' *Business and Society Review* 75, 4–5

Shimell, Pamela (1991) 'Corporate environmental policy in practice' *Long Range Planning* 24, 3, 10–17

Simmons, Peter (1993) 'Greening consumers? Environment and politics in the marketplace' presented at IRNES Conference: Perspectives on the Environment 2, Sheffield University, UK

Simmons, Peter and John Cowell (1993) 'Liability for the environment: lessons from the development of civil liability in Europe' 345–64 in Tim Jackson (ed.) *Clean Production Strategies: Developing Preventive Environmental Management in the Industrial Economy* (Lewis Publishers, London)

Simmons, Peter and Brian Wynne (1993) 'Responsible Care: trust, credibility and environmental management' 201–26 in Kurt Fischer and Johan Schot (eds) *Environmental Strategies for Industry: International Perspectives on Research Needs and Policy Implications* (Island Press, Washington DC)

Simms, Christine (1992) 'Green issues and strategic management in the grocery retail sector' *International Journal of Retail and Distribution Management* 20, 1, 32–42

Sinclair, John (1987) *Images Incorporated: Advertising as Industry and Ideology* (Croom Helm, London)

Smart, Bruce (1992) *Beyond Compliance: A New Industry View of the Environment* (World Resources Institute, London)

Smith, Charlotte and Clare Sambrook (1990) 'Dead End Street' *Marketing* May 10, 28–31

Smith, Denis (1993a) 'Business and the environment: towards a paradigm shift?' 1–11 in Denis Smith (ed.) *Business and the Environment: Implications of the New Environmentalism* (Paul Chapman, London)

Smith, Denis (1993b) 'The Frankenstein syndrome: corporate responsibility and the environment' 172–89 in Denis Smith (ed.) *Business and the Environment: Implications of the New Environmentalism* (Paul Chapman, London)

Smith, N Craig (1990) *Morality and the Market: Consumer Pressure for Corporate Accountability* (Routledge, London)

Spretnak, Charlene and Fritjof Capra (1985) *Green Politics* (Paladin, London)

Steger, Ulrich (1993) 'The greening of the board room: how German companies are dealing with environmental issues' 147–66 in Kurt Fischer and Johan Schot (eds) *Environmental Strategies for Industry: International Perspectives on Research Needs and Policy Implications* (Island Press, Washington DC)

Streeck, Wolfgang and Philippe C Schmitter (eds) (1985) *Private Interest Government: Beyond Market and State* (Sage, London)

Stroup, Margaret A, Ralph L Neubert and Jerry W Anderson Jr (1987) 'Doing good, doing better: two views of social responsibility' *Business Horizons* March/April 22–5

Tanega, Joseph (1993) 'Towards an environmental credit-rating agency' 107–16 in Jane Holder, Pauline Lane, Sally Eden, Rachel Reeve, Ute Collier and Kevin Anderson (eds.) *Perspectives on the Environment: Interdisciplinary Research in Action* (Avebury, Aldershot)

Taylor, Stuart R (1992) 'Green management: the next competitive weapon' *Futures* September 669–80

Touche Ross (1990) *Head in the Clouds or Head in the Sand? UK Managers' Attitudes to Environmental Issues: A Survey* (Touche Ross Management Consultants, London)

Trippier, David (1990) 'Introduction' 7–11 in CBI/ICC *Environmental Auditing* Conference transcript (Rooster, Royston, Herts)

Turner, Kerry (ed.) (1988) *Sustainable Environmental Management: Principles and Practice* (Belhaven, London)

UNEP (UN Environment Programme)/IEO (Industry and Environment Office) (1991) *Companies' Organization and Public Communication on Environmental Issues* (UNEP/IEO, Paris)

Useem, Michael (1984) 'Business and politics in the United States and United Kingdom' 263–91 in Sharon Zukin and Paul DiMaggio (eds) *Structures of Capital: The Social Organization of the Economy* (Cambridge University Press, Cambridge)

Van Liere, Kent D and Riley E Dunlap (1981) 'Environmental concern: does it make a difference how it's measured?' *Environment and Behaviour* **6**, 651–76

Verheul, Hugo and Philip J Vergragt (1994) 'Demand articulation by citizen-led initiatives' presented at the Greening of Industry Conference, Copenhagen, Denmark

Verlander, Neil (1992) 'Pressure Group Perspective: Friends of the Earth "Green Con of the Year Award"' in Martin Charter (ed.) *Greener Marketing: A Responsible Approach to Business* (Greenleaf Publishing, Sheffield)

Vogel, David (1986) *National Styles of Regulation: Environmental Policy in Great Britain and the United States* (Cornell University Press, Ithaca, NY)

WCED (1987) *Our Common Future* (Oxford University Press, Oxford)

Welford, Richard (1993) 'Breaking the link between quality and the environment: auditing for sustainability and life cycle assessment' *Business Strategy and the Environment* **2**, 4, 25–33

Welford, Richard (1994) *Environmental Strategy and Sustainable Development: The Corporate Challenge for the 21st Century* (Routledge, London)

Welford, Richard and Andrew Gouldson (1993) *Environmental Management and Business Strategy* (Pitman, London)

Wernick, Andrew (1991) *Promotional Culture: Advertising, Ideology and Symbolic Expression* (Sage, London)

West, Karen (1995) 'Ecolabels: the industrialization of environmental standards' *The Ecologist* **25**, 1, 16–20

WICE (1994) *Environmental Reporting: A Manager's Guide* (WICE, Paris)

Williams, Marc (1993) 'International trade and the environment: issues, perspectives and challenges' *Environmental Politics* **2**, 4, 80–97

Williams, Raymond (1980) *Problems in Materialism and Culture: Selected Essays* (Verso, London)

Williamson, Judith (1978) *Decoding Advertisements: Ideology and Meaning in Advertising* (Marion Boyars, London)

Williamson, Peter (1989) *Corporatism in Perspective: An Introductory Guide to Corporatist Theory* (Sage, London)

Willums, Jan-Olaf and Ulrich Golüke (1992) *From Ideas to Action: Business and Sustainable Development. The ICC Report on the Greening of Enterprise '92* (ICC/Ad Notam Gyldenal, Oslo)

Wilson, Alexander (1992) *The Culture of Nature: North American Landscape from Disney to the Exxon Valdez* (Blackwell, Oxford)

Wilson, Frank L (1990) 'Neo-corporatism and the rise of new social movements' 68–83 in Russell J Dalton and Manfred Kuechler (eds.) *Challenging the Political Order: New Social and Political Movements in Western Democracies* (Polity, Cambridge)

Wohl, Anthony S (1983) *Endangered Lives: Public Health in Victorian Britain* (J M Dent, London)

Worster, Donald (1985) *Nature's Economy: A History of Ecological Ideas* (Cambridge University Press, Cambridge)

Yearley, Steven (1991) *The Green Case: A Sociology of Environmental Issues, Arguments and Politics* (Harper Collins, London)

Zucker, Lynne G (1986) 'The production of trust: institutional sources of economic structure 1840–1920' *Research in Organizational Behaviour* 8, 53–111

Index